Essential Chemistry for
FORMULATORS OF SEMISOLID AND LIQUID DOSAGES

Essential Chemistry for
FORMULATORS OF SEMISOLID AND LIQUID DOSAGES

Authors

VITTHAL S. KULKARNI, Ph.D.
Scientific Advisor, Research and Development, DPT
Laboratories, Ltd., San Antonio, TX, USA

CHARLES SHAW, Ph.D.
Scientific Advisor, Research and Development, DPT
Laboratories, Ltd., San Antonio, TX, USA

ELSEVIER

Amsterdam • Boston • Heidelberg • London
New York • Oxford • Paris • San Diego
San Francisco • Singapore • Sydney • Tokyo
Academic Press is an imprint of Elsevier

Academic Press is an imprint of Elsevier
125 London Wall, London EC2Y 5AS, UK
525 B Street, Suite 1800, San Diego, CA 92101-4495, USA
225 Wyman Street, Waltham, MA 02451, USA
The Boulevard, Langford Lane, Kidlington, Oxford OX5 1GB, UK

Notices
Knowledge and best practice in this field are constantly changing. As new research and
experience broaden our understanding, changes in research methods, professional practices,
or medical treatment may become necessary.

Practitioners and researchers must always rely on their own experience and knowledge in
evaluating and using any information, methods, compounds, or experiments described
herein. In using such information or methods they should be mindful of their own safety
and the safety of others, including parties for whom they have a professional responsibility.

To the fullest extent of the law, neither the Publisher nor the authors, contributors, or
editors, assume any liability for any injury and/or damage to persons or property as a
matter of products liability, negligence or otherwise, or from any use or operation of any
methods, products, instructions, or ideas contained in the material herein.

ISBN: 978-0-12-801024-2

British Library Cataloguing-in-Publication Data
A catalogue record for this book is available from the British Library

Library of Congress Cataloging-in-Publication Data
A catalog record for this book is available from the Library of Congress

For information on all Academic Press publications
visit our website at http://store.elsevier.com/

Working together
to grow libraries in
developing countries

www.elsevier.com • www.bookaid.org

Publisher: Mica Haley
Acquisition Editor: Kristine Jones
Editorial Project Manager: Molly McLaughlin
Production Project Manager: Karen East and Kirsty Halterman
Designer: Greg Harris

Typeset by TNQ Books and Journals
www.tnq.co.in

Printed and bound in the United States of America

CONTENTS

PREFACE

The focus of this book is the formulation, product development, testing, and stability of semisolid and liquid dosage forms. Formulations of this type are important throughout the industry, and are an essential vehicle for delivering pharmaceuticals safely and effectively to their intended target. Understanding the chemistry of surfactants and surface phenomena is key to successfully formulating complex multicomponent formulations.

The formulation of semisolid dosages requires mixing several ingredients to produce a uniform, homogeneous, stable product that can be filled into suitable containers for the end user. This is also applicable to some liquid formulations that are multicomponent systems. The mixing of various ingredients that may be mutually immiscible liquids containing partially soluble or insoluble materials to form a uniform, stable system requires a thorough knowledge of chemistry. The required knowledge of physical, organic, and analytical chemistry that is essential for pharmaceutical formulation development is well documented in various books and research publications. The purpose of this book is not to give fundamental details of the chemistry, but to draw attention to the chemical principles underlying the formulation development of complex mixtures. As some of the medications will be delivered through specialized devices, a basic knowledge of engineering and, to a certain extent, physics is also beneficial. Furthermore, an important aspect for any chemist or pharmacist involved in pharmaceutical drug product development is an awareness of the regulatory agency's requirements, generally known as the "Chemistry, Manufacturing, and Controls" part of a drug product regulatory filing. In this book we have focused on surfactants, thickeners, surface chemistry, drug delivery systems, methods used for characterization, and regulatory aspects of testing drug products.

We hope that this book will be helpful to scientists involved in the drug product development of semisolid and liquid dosage forms, and that it will give newcomers to this field a broad perspective of the chemistry involved in formulating and testing drug products.

The process of writing and reviewing chapters was very time-consuming and took many hours of personal time away from our families. We are thankful to our colleagues and friends who assisted us at various stages. We remain indebted to our wives Anuradha and Karen, and our family members who supported us throughout the long process.

Vitthal S. Kulkarni and Charles Shaw

September 2015

CHAPTER 1

Introduction

Every year, several new drug products based on either new drug substances or generics of existing drug products are approved and enter the market. In 2014, approximately 95 Abbreviated New Drug Applications (ANDAs, i.e., generic) and 106 New Drug Applications (NDAs) were approved by the Center for Drug Evaluation and Research (CDER), a division of the US Food and Drug Administration (FDA). This demonstrates that Research and Development within the pharmaceutical industry is a very competitive field. Achieving success in drug formulation development requires a combined knowledge of chemistry, chemical and process engineering, and the regulations. Drug product development activities include aspects of preformulation/formulation development (including compatibility of the API with formulation excipients, and the compatibility of the ingredients and finished formulation with the manufacturing process, process parts, and container–closure system), as well as aspects of manufacturing and stability testing.

Due to concerns relating to toxicity and possible side effects, very few drug substances can be directly administered to the body. Additionally, the amount of the drug substance administered (i.e., the maximum daily dose) is often in milligram or microgram quantities. As a result, it is necessary to mix the drug with other nondrug (inactive) ingredients in such a way that the drugs can be safely delivered to their target within the body.

The US FDA definitions for active ingredient, drug, and drug products are:

- Active Ingredient: any component that provides pharmacological activity or other direct effect in the diagnosis, cure, mitigation, treatment, or prevention of disease or affects the structure or any function of the body of man or animal.
- Drug:
  - A substance recognized by an official pharmacopeia or formulary.
  - A substance intended for use in diagnosis, cure, mitigation, treatment, or prevention of disease.

Essential Chemistry for Formulators of Semisolid and Liquid Dosages
http://dx.doi.org/10.1016/B978-0-12-801024-2.00001-7

- A substance (other than food) intended to affect the structure or any function of the body.
- A substance intended for use as a component of medicine but not a device or a component, part, or accessory of a device.

Biological products are included within this definition and are generally covered by the same laws and regulations, but differences exist regarding their manufacturing process.

- Drug Product: a finished dosage form that contains an active drug ingredient, generally, but not necessarily, in association with other active or inactive ingredients.

As an example, a tablet of Ibuprofen of 200 mg strength may actually weigh 500 mg, indicating that to deliver 200 mg of the drug substance, 300 mg of inactive ingredients are added to help safely deliver the drug to its destination. The added inactive ingredients might also help to keep the drug stable for a finite period. Therefore, converting a drug substance into a safe and effective drug product is equally as important as inventing the drug itself. This process of making drug products by combining the drug substance (the API) and the necessary inactive ingredients (the excipients) is called the formulation process. When formulating a drug product, the goal is to make a safe and effective product that has an acceptable shelf life, which can be easily administered (to promote patient compliance).

Pharmaceutical formulations are mixtures of the pharmaceutically active ingredient and selected inactive ingredients. Solution formulations that are used for injectable dosage forms generally have fewer inactive ingredients—such as water, cosolvents, buffering agents, and pH-adjusting agents. As a result, they are much simpler to formulate compared to some of the semisolid formulations used for topical administration. The inactive ingredients used in semisolid formulations may include water, oil, surfactants, emulsifiers, stabilizers, chelators, preservatives, and pH-adjusting agents. These types of formulations tend to be complex due to interactions between the various ingredients and consequently require considerable development efforts during the formulation process. Furthermore, when formulating a generic version of an existing marketed product, reverse engineering of the reference drug product is often challenging for semisolids.

For all types of formulations, product development efforts require putting together all of the ingredients, testing their mutual compatibility (e.g., the drug substance with the inactive ingredients, and between one inactive ingredient and another), the solubility of the API in the formulation matrix, the chemical stability of each ingredient, and the physical stability of the formulation as a whole. Developing stability-indicating test methods is an inherent part of product development activities. The US Food and Drug Administration (FDA) recommend using a Quality by Design (QbD) approach when developing drug products, including generic products. When developing generics, three critical attributes are (1) the ingredients are the same as in the reference-listed drug (RLD); (2) the concentrations of the ingredients are the same as in the RLD; and (3) the microstructure (arrangement of matter) within the generic is the same as in the RLD.

Common types of dosage forms are solid, liquid, aerosol, suspension, and semisolid. Solids are typically administered either orally, by inhalation (e.g., dry powder inhalers), or applied topically (e.g., powders). Liquids can be injected, taken orally, applied topically, or administered via the nasal or pulmonary route (in the form of a spray). Semisolids can be administered topically (e.g., creams, lotions, ointments, gels), transdermally, injected subcutaneously (e.g., gels for subcutaneous administration), vaginally, or anally (e.g., suppositories).

In recent years, several medications have appeared on the market in which the drug product is delivered via a device in which the device is an integral part of the medication. The term "Combination Products" is used when the medication is composed of any combination of drug and a device. Examples of combination products include pressurized metered-dose inhalers, nebulizers, dry-powder inhalers, transdermal patches that deliver drugs by iontophoresis or microneedle technology, and prefilled syringes. The definition of a combination product includes:

- A product comprising two or more regulated components (i.e., drug/device, biologic/device, drug/biologic, or drug/device/biologic) that are physically, chemically, or otherwise combined or mixed and produced as a single entity.
- Two or more separate products packaged together in a single package or as a unit and comprising drug and device products, device and biological products, or biological and drug products.
- A drug, device, or biological product packaged separately that, according to the investigational plan or proposed labeling, is intended for use only with an approved individually specified drug, device, or biological product in which both are required to achieve the intended use, indication, or effect and in which upon approval of the proposed product the labeling of the approved product would need to be changed.

- Any investigational drug, device, or biological product packaged separately that according to its proposed labeling is for use only with another individually specified investigational drug, device, or biological product in which both are required to achieve the intended use, indication, or effect.

Inactive ingredients that act in the formulation as surfactants, emulsifiers, thickeners, gelling agents, or stabilizers play critical roles in making stable semisolid formulations. To successfully design and develop stable drug products, it is critical to have a knowledge and understanding of surfactants and surface chemistry. Formulation development is not complete until the various ingredient compatibility aspects are studied and data are produced. These compatibility aspects include API and excipients, solubility of the API in the formulation matrix, and compatibility of the formulation with the intended primary container-closure system. If more than one API is to be incorporated in the formulation, the process becomes more complicated. Establishing processing compatibility at all stages of the product development life cycle is also a key aspect of formulation development—including scale-up, full-scale manufacture, and the filling process. To assess the risks and take appropriate measures to reduce them, the FDA recommends following a QbD approach for all stages of formulation and process development. A detailed discussion surrounding QbD and stability testing is provided in this book.

Chapters covering the rheology, particle size, microscopy, and miscellaneous physical and chemical test methods for semisolid and liquid dosage forms are included. Formulation topics such as the use of polymers, thickeners, other excipients, and methods of manufacturing are also covered. A chapter is also devoted to aerosol formulations, which is a rapidly growing area in terms of new products.

Finally, no drug product can be placed into the market without approval by regulatory agencies. Satisfying the regulatory requirements for new drug products or generics is essential. A full chapter is devoted to a review of the current regulatory landscape relating to semisolid and liquid dosage forms.

CHAPTER 2

Surfactants, Lipids, and Surface Chemistry

Contents

2.1 INTRODUCTION

This chapter presents an introduction to surface chemistry and surfactants in pharmaceutical formulations. The choice and quantity of surfactants are essential factors in the formation of stable and efficacious emulsions as pharmaceutical dosage forms (including oral, topical, and injectable/infusible dosage forms). Understanding the physical chemistry of surfactant molecules in aqueous systems and at the air–water or liquid–liquid interface is fundamental to designing drug delivery systems. Important parameters include surface tension, contact angle, and critical micelle concentration.

Surfactants are "amphiphilic" molecules (or "amphiphiles"), containing both hydrophilic and hydrophobic portions on the same molecule. They have been termed "surface active agents" due to their activity at the air–water or water–oil interfaces, reducing surface tension and improving miscibility. They are also

Essential Chemistry for Formulators of Semisolid and Liquid Dosages
http://dx.doi.org/10.1016/B978-0-12-801024-2.00002-9

Figure 2.1 Schematic representation of a surfactant molecule. The circle represents the hydrophilic part of the molecule (also known as the head group or polar group), and the long acyl chain represents the hydrophobic (lipophilic) part of the molecule.

considered as chemicals that form "new surfaces" (micelles, liposomes, etc.) in the presence of water [1]. Surfactants are a class of chemicals that has ubiquitous applications in industries; almost all industries need to use surfactants, including engineering, food, petrochemical, pharmaceutical, and consumer products.

2.2 TYPES OF SURFACTANTS

A schematic representation of a surfactant molecule is shown in Figure 2.1; the hydrophilic ("water loving") end of the surfactant molecule is referred to as the "head group" or polar group, and the hydrophobic ("water hating") portion is referred to as the "tail" (or lipophilic, oil-soluble end). Depending on the charge of the polar group, surfactants are generally classified as cationic, anionic, nonionic, or zwitterionic (amphoteric) surfactants. Surfactants can be water soluble or insoluble, natural or synthetic in origin, and their chemical structures could be simple or large molecules or polymeric. Surfactants that have polar groups at each end of a long acyl chain are sometimes known as "bolaamphiphiles" [2].

When mixed in water, surfactants reduce the surface tension of the water. As the concentration of the surfactant is increased, the surface tension continues to drop. Above a certain concentration, the surfactant molecules spontaneously form micelles. When these micelles start forming, further additions of surfactant have no further effect on the surface tension of the water. This concentration, at which the surface tension remains constant, is called the Critical Micelle Concentration (CMC). A typical representation of surface tension versus surfactant concentration is shown in Figure 2.2.

2.2.1 Surfactant Micelles

Micelles are self-assembled microstructures formed by surfactants in aqueous systems. They can trap hydrophobic molecules in their hydrophobic core and thereby act as wetting agents or solubilizing agents. This behavior enables them to be effective cleansing agents. Depending on the structure of the surfactant molecules, micelles of different shapes (including spherical, cylindrical, hexagonal, cubic lamellar, inverted cylindrical, and inverted spherical) can be formed.

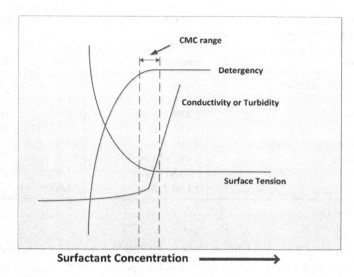

Figure 2.2 A schematic representation of trend lines for surfactant physical properties such as surface tension, conductivity, turbidity, and detergency as a function of surfactant concentration. A change in trend is observed after Critical Micelle Concentration (CMC) is reached. *Adapted from Barnes, G. and Gentle, I., Interfacial Science An Introduction, Oxford University Press, 2005.*

The size of a micelle is related to the number of monomers per micelle (commonly called the aggregation number) or the molecular weight of the micelle.

Factors affecting the formation of micelles include:

- Chain length of the surfactant molecule: molecules with longer chain lengths are less soluble, and will form micelles at lower concentrations.
- The CMC of ionic surfactants is greatly affected by the presence of dissolved salts in the solution. The CMC is lowered as the concentration of salts is increased [3,4] (see Table 2.1).
- When alcohol is added to the water, the CMC increases considerably. It has been shown that, within a series of alcohols ranging from methanol to butanol (at 10% concentration in water), there was no uniform trend observed for the change in CMC for polysorbate 20 [5].
- Increase in temperature increases the CMC.
- If two or more surfactants are present in a solution, the surfactant with higher CMC acts like an electrolyte, and the overall CMC of the mixture will depend on the nature of the individual surfactants (i.e., ionic or nonionic).

2.2.2 Anionic Surfactants

Surfactants with a negative charge on the head group are called "anionic." Common examples are sodium salts of fatty acids and fatty sulfates, including

Table 2.1 Change in critical micelle concentration (CMC) and aggregation number (N) with salt concentration for ionic surfactants

Surfactant	Medium	CMC (mM)	N
Sodium dodecylsulfate	Water	8.1	58
	0.1 M NaCl	1.4	91
	0.2 M NaCl	0.83	105
	0.4 M NaCl	0.52	129
Dodecyltrimethylammonium bromide	Water	14.8	43
	0.0175 M NaBr	10.4	71
	0.05 M NaBr	7.0	76
	0.1 M NaBr	4.65	78

Data from Ref. [3].

sodium lauryl sulfate. Typically, sodium, potassium, or ammonium salts of fatty sulfonates with C8–C14 acyl chains tend to possess foaming and cleansing properties, and are suitable for use in shampoo-type applications. Anionic surfactants also assist in the dissolution or absorption of drugs. Bile salts in the stomach are anionic surfactants and play a critical role in food digestion [6,7].

2.2.3 Cationic Surfactants

Surfactants with a positive charge on their head group are called "cationic." Examples of cationic surfactants include long-chain quaternary ammonium compounds (e.g., Dimethyldioctadecylammonium chloride). Some long-chain quaternary surfactants show antimicrobial properties, and are used as preservatives in pharmaceutical formulations.

2.2.4 Nonionic Surfactants

These do not ionize in the presence of water, and generally have a low irritancy potential compared to ionic surfactants. The hydrophilic portion of the nonionic surfactant molecule could be an alcohol, polyol derivative, or esters of fatty molecules (waxes). Nonionic surfactants are widely used in pharmaceutical formulations.

The structure of some of the synthetic surfactants are shown in Figure 2.3.

2.3 NATURAL SURFACTANTS

Surfactants are abundant in nature, and are present in both plants and animals. Some proteins behave like surfactants and play critical roles in various biological processes. An important group of protein surfactants

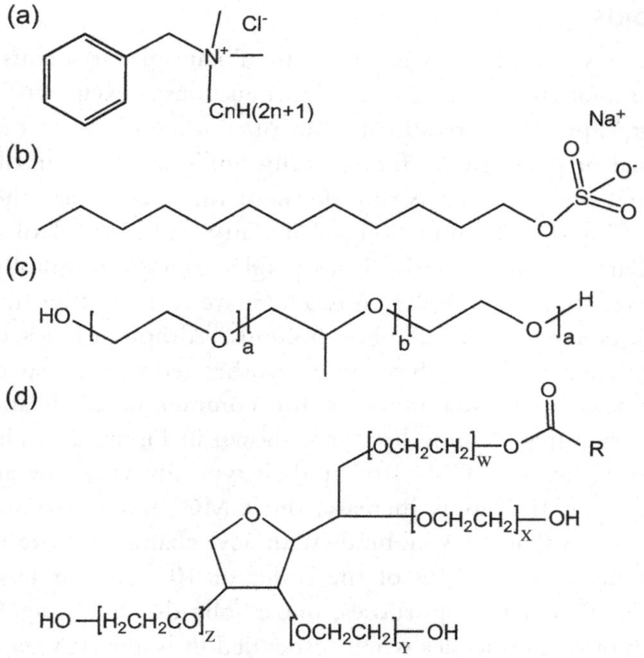

Figure 2.3 (a) Benzalkonium chloride where n = C_8 to C_{18}. (b) Sodium lauryl sulfate (sodium dodecyl sulfate). (c) Poloxamer (for Poloxamer 124, a = 12 and b = 20; Poloxamer 188, a = 80 and b = 27; Poloxamer 237, a = 64 and b = 37; Poloxamer 338, a = 141 and b = 44; Poloxamer 407, a = 101 and b = 56). (d) Polysorbates (W + X + Y + Z = 20); For Polysorbates 20, 40, 60, and 80, R = laurate, palmitate, stearate, and oleate, respectively.

from a therapeutic point of view are "lung surfactants" (or "pulmonary surfacants") [8]. In humans, lung surfactants are composed of approximately 90% lipids (phospholipids, cholesterol, and other lipids) and four types of pulmonary surfactant proteins (SP-A, SP-B, SP-C, and SP-D). The lung surfactants facilitate gas exchange in the lungs and prevent them from collapsing. The deficiency of lung surfactants can be caused by several reasons, including: premature birth, lung injury, or genetic mutations that inhibit surfactant production or function. Patients with lung surfactant deficiency are in risk of respiratory failure (e.g., Respiratory Distress Syndrome or Acute Respiratory Distress Syndrome) and the condition needs to be treated using lung surfactant supplements. Examples of medications include Survanta® and Surfaxin® (visit http://reference.medsca pe.com/drugs/lung-surfactants for additional information). Other examples of natural surfactants are bile salts [6,7,9].

2.3.1 Lipids

Lipids are a subset of a large domain of natural surfactants and are extremely important in pharmaceutical formulations as excipients. The word "lipid" (or "lipide" in French) implies organic compounds originating from animal or plant grease/fat. Typically, lipids are water insoluble and their natural physical state is oily or waxy. Included under the general category of "lipids" are simple long-chain fatty acids, fatty alcohols, esters, amines, plant- or animal-derived oils (triglycerides), phospholipids (also known as lecithin), and cholesterols. There are several other lipid classes within the general group of "phospholipids." Although lipids of natural origin are water insoluble, they can be synthesized to meet specific head group or acyl chain requirements for commercial applications. The chemical structures of some lipids are shown in Figure 2.4. The critical micelle concentration (CMC) of lipids is typically very low and as the chain length of the lipids increase, the CMC decreases considerably. Table 2.2 shows that phospholipids with acyl chains of more than nine carbon atoms have a CMC of the order of 10^{-3} mM or less. Consequently, for all practical purposes, phospholipids with long fatty acyl chains do not form micelles when suspended in water. At very low concentrations they form monomolecular films at the air–water interface. When in excess, they self-assemble into bilayers that are the building blocks of cell membranes.

Lipids play a very significant role in pharmaceutical formulations of parenteral, topical, transdermal, and oral dosage forms [10]. The lipids not only form drug delivery systems for both water-soluble and water-insoluble drugs, they also enhance drug penetration through the skin and drug absorption through intestinal mucosal membranes for oral dosages. Lipid analysis of human skin has been extensively investigated, and it is well established that a variety of lipids including fatty acids, phospholipids, ceramides, and many other types are present [11,12]. Topical or transdermal formulations containing lipids are considered to be well tolerated/less irritating to human skin [13,14].

Lipid-based formulations (LBS) of oral dosages, in particular liquid formulations in soft-gel capsules, have shown improved efficacy by enhancing the bioavailability of the drug. Digestion and dispersion are critical factors that affect the performance of oral dosages of lipid-based formulations. This is because during intestinal processing, physicochemical properties may change and impact performance by causing drug

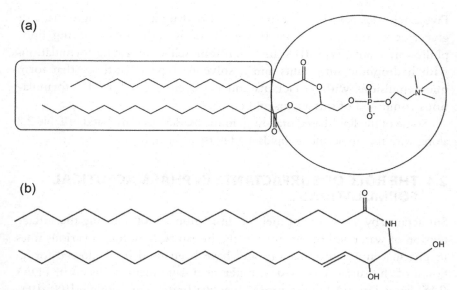

Figure 2.4 (a) Structure of dipalmitoyl phosphatidyl choline (DPPC), a typical phospholipid. *(From Kulkarni VS, Liposomes in personal care products. Delivery system handbook for personal care and cosmetic products. In: Meyer R. Rosen, editor. Technology, applications and formulations; in print 2005).* (b) A type of sphingo lipid known as ceramide; several different types of ceramides (variations in N-acyl chain length) are present in stratum corneum of human skin.

Table 2.2 Critical micelle concentrations for lecithins with increasing acyl chain lengths

Lipids	CMC (mM)
Dibutanoyl lecithn	80
Dihexanoyl lecithin	14.6
Dioctanoyl lecithin	0.265
Dinonanoyllecithin	2.87×10^{-3}
Dipalmitoyl phosphatidyl choline	2×10^{-8}

Compiled from Refs [31].

precipitation before the drug is absorbed. In order to aid formulation development and assess which factors affect performance, standardized in vitro methods have been developed along with a system for classifying LBS formulations [15–17]. This classification is based on formulation components and their dependence on digestion to facilitate dispersion.

Five categories have been proposed; Type-I: drug mixed in triglyceride or glyceride formulations; Type-II: formulations additionally having lipophilic surfactants; Type-IIIA: fine emulsion–self-emulsifying formulations with hydrophilic surfactants and cosolvents; Type-IIIB: those that form microemulsions with surfactants and cosolvents; and Type-IV: formulations composed of surfactants and cosolvents.

Some of the lipid-based oral drugs in the marketplace are listed in Table 2.3 along with the surfactants or lipids used in their formulations.

2.4 THE ROLE OF SURFACTANTS IN PHARMACEUTICAL FORMULATIONS

Surfactants, by the virtue of their intrinsic property of reducing the surface tension of water and being amphiphilic in nature, have found various roles in pharmaceutical formulations and have been used in all dosage forms. Some of the surfactants used in different dosage forms as listed in FDA's (U.S. Food Drug Administration) "Inactive Ingredient Guide" (IIG) database are shown in Table 2.4.

2.4.1 Skin Penetration Enhancers

The primary function of the skin is to protect the internal organs from external invasion by forming a strong barrier between the outside environment and the body. However, for topical dosage formulations such as transdermals, penetration of the drug into the skin is essential. Therefore, transdermal/topical formulations need to act against the natural function of the skin [18]. Although the skin acts as a barrier, it is not completely impermeable. Chemical penetration enhancers and certain surfactants can be incorporated into topical/transdermal formulations to help facilitate penetration of the drug into the skin [19]. Fatty acids (e.g., oleic acid), fatty alcohols (e.g., myristyl or oleoyl alcohols), fatty esters (e.g., isopropyl myristate), lipids, anionic and nonionic surfactants are commonly used as chemical penetration enhancers. Although the use of surfactants can enhance penetration of the drug into the skin, they increase the risk of skin irritation. Consequently, skin irritation potential becomes a critical factor to be considered when formulating products for use on compromised skin. Formulations targeted for mucosal membranes (such as nasal, buccal cavity, vaginal, or suppositories) generally use nonionic surfactants as they are less irritating than ionic surfactants.

Table 2.3 Lipids and surfactants used in some drug products in the marketplace

Trade names	Drug	Therapeutic use	Lipids/surfactants
Agenerases	Amprenavir	HIV antiviral	Tocopherol PEG succinate, and PEG 400
Rocaltrols	Calcitriol	Calcium regulator	Medium-chain triglycerides
Sandim-mune	Cyclosporin	Immuno-suppressant	Corn oil, and linoleoyl macrogolglyceride
Neorals	Cyclosporin A/I	Immuno-suppressant	Corn oil monodiglycerides, and polyoxyl 40 hydrogenated castor oil
Gengrafs	CyclosporinA/III	Immuno-suppressant	Polyoxyl 35 castor oil, polysorbate 80, and sorbitan monooleate
Accutanes	Isotretinoin	Anti-comedo-genic	Hydrogenated soybean oil flakes, hydrogenated vegetable oils, and soybean oil
Kaletras	Lopinavir and ritonavir	HIV antiviral	Oleic acid, and polyoxyl 35 castor oil
Norvirs	Ritonavir	HIV antiviral	Oleic acid, and polyoxyl 35 castor oil
Lamprene	Clofazamine	Treatment of leprosy	Rapeseed oil, soybean lecithin, hydrogenated soybean oil, and partially hydrogenated vegetable oils
Sustivas	Efavirenz	HIV antiviral	Sodium lauryl sulfate, and sodium starch glycolate
Lofibra	Finofibrate	Lipid-regulating	Sodium lauryl sulfate
Restandol	Testosteroneun-decanoate	Hormone replacement therapy	Castor oil, and propylene glycol laurate
Prometrium	Progesterone	Hyperplasia	Peanut oil, and lecithin
Rapamune	Sirolimus	Immuno-suppressant	Phosphatidylcholine, and polysorbate 80
Vesanoid	Tretinoin	Acute promy-elocytic leukemia	Hydrogenated soybean oil flakes, hydrogenated vegetable oils, and soybean oil

Compiled from Refs [15,16].

Table 2.4 Some of the surfactants listed in FDA's Inactive ingredient guide with the reported use in different dosage forms

Type of surfactant	Surfactant	Dosage form
Anionics	Ammonium lauryl sulfate	Topical aerosol, emulsions
	Sodium lauryl sulfate	Topical creams, gels, oral capsule, drops, tablets
	Alkyl aryl sodium sulfonate	Topical suspension
	Sodium cholestryl sulfate	IV (infusion), injection suspensions
Catatonics	Polyquaternium 10	Topical
	Quaternium-15 and -52	Topical emulsion, aerosol foams
	Benzalkonium chloride	Topical, nasal, ophthalmic, injectables
	Cetrimonium chloride	Topical
	Cetylpyridinium chloride	Inhalation aerosol, capsules
Nonionics	Polysorbate X (X = 20 or 40 or 60 or 65 or 80)	Injection, IV infusion, nasal sprays, ophthalmic suspensions, topical creams
	Sorbitan monolaurate (or sorbitan monostearate)	Topical lotions, emulsions ointments
	Glyceryl oleate (monostearate or dibehenate)	Topical, oral
	Glyceryl trioleate	Injection suspensions
	PEG 5 oleate	Topical, vaginal
	Poloxamers (e.g., 124, 188, 407)	Oral, topical, I.V. injection, subcutaneous, and ophthalmic
	Polyglyceryl-3-oleate	Oral capsules
Zwitterionics	Dioleoyl glycerophosphocholines	Injections, liposomal
	Egg phospholipids	Intravenous, injection
	Lecithin (soy lecithin)	Liposomal, injection, oral, nasal, inhalation aerosol
	Coco betaine	Topical

Compiled from http://www.accessdata.fda.gov/scripts/cder/iig/index.cfm.

2.4.2 Emulsifying Agents

Oil and water are immiscible with each other because of the very high interfacial tension at the oil–water interface. In the presence of surfactants, however, this oil–water interfacial tension can be reduced to such an extent that the two immiscible phases can be made miscible. Many topical/transdermal dosages are creams that are emulsions of oil and water phases. Other emulsion dosage forms including injectables, nasal sprays, and ophthalmic drops are available commercially. The selection of the surfactants to use and their concentration in the formulation are critical to making stable emulsions. A study of the oil–water interfacial tension as a function of surfactant concentration helps to determine the critical concentration of surfactant needed to achieve emulsification.

2.4.2.1 Hydrophile–Lipophile Balance (HLB) System

Surfactants are used to produce emulsions of oil droplets dispersed in water (oil-in-water) or water droplets dispersed in oil (water-in-oil). Consequently, emulsions have a large technological application, including in pharmaceutical dosages. An empirical but very useful numerical rating system was introduced by Griffin and is known as the Hydrophile–Lipophile Balance or HLB number [20]. In general, highly water-soluble surfactants have high HLB values and highly oil-soluble surfactants have low HLB values. HLB number ranges based on the solubility or dispersability of the surfactants are shown in Table 2.5.

Generally, surfactants with an HLB in the range of 4–6 are water-in-oil emulsifiers and 8–18 are oil-in-water emulsifiers. HLB values can also be determined experimentally as $HLB = 20(1 - S/A)$ in which S = saponification number of the ester and A = acid value of recovered acid.

HLB numbers for some common surfactants are shown in Table 2.6.

2.4.3 Aerosol Formulations

Surfactants have been used extensively in aerosol formulations. Surfactants reduce the surface tension of water and thereby facilitate atomization of the formulation. For nasal spray formulations, the formation of uniform plumes depends not only on the device but also on the formulation. The use of surfactants in nasal sprays is common to achieve the effective delivered dose (proper droplet size and plume) [22]. The use of surfactants in topical aerosols (wound healing sprays or pain relief sprays) or foam formulations is also common.

2.4.4 Surfactant Gels

The formation of gels by small amphiphilic molecules (surfactants) is well documented [23–25]. Gel formation by surfactants is considered a process similar to micellization rather than crystallization. Poloxamers™ are polymeric nonionic surfactants of ethylene oxide and propylene oxide, and several varieties of Poloxamers™ are listed in the FDA IIG database for use in pharmaceutical

Table 2.5 Surfactant water solubility and HLB ranges

Water solubility	HLB range
No dispersibility in water	1–4
Poor dispersion	3–6
Milky dispersion after vigorous shaking	6–8
Stable milky dispersion	8–10
Translucent to clear dispersion	10–13
Clear solution	13+

Table 2.6 HLB values for some common surfactants

Surfactant	HLB
Sorbitan tristearate	2.1
Sucrose distearate	3
Glyceryl monooleate	3.4
Glyceryl monostearate	3.8
Span 80 (sorbitan monooleate)	4.3
Glyceryl monolaurate	5.2
Sorbitan monopalmitate	6.7
Soy lecithin	8
Sorbitan monolaurate	8.6
Tween 81	10
Tween 80	15
Sodium stearoyl-2-lactylate	12
Sodium oleate	18
Ammonium lauryl sulfate	31

Some of the values are from Ref. [21].

formulations. Two of the Poloxamers™, Poloxamer 188 and Poloxamer 407, exhibit thermo-sensitive properties—that is, they are soluble in water at low temperature and gel at higher temperature.

2.5 SURFACE CHEMISTRY FOR PHARMACEUTICAL FORMULATIONS

2.5.1 Surface and Interfacial Tension

Physical properties including surface tension, osmotic pressure, conductivity, and detergency will change (either increase or decrease) as the concentration of surfactant increases. There are several methods, including Du Nouy ring, Wilhelmy plate, maximum bubble pressure, drop volume, pendant drop, sessile drop, spinning drop, and capillary rise available to measure surface and interfacial tension [26]. Du Nouy ring and Wilhelmy plate can be used for both air–liquid and liquid–liquid interfacial tension measurements, and these are the most commonly used methods. American Society for Testing and Materials method number D1331-14 describes the test methods for "Surface and Interfacial Tension of Solutions of Paints, Solvents, Solutions of Surface-Active Agents, and Related Materials." Measurements of interfacial tension between oil and water in the presence of a surfactant provide information on the amount of surfactants required to reduce the interfacial tension to a level at which an emulsion can be formed. Pecora et al. [27] have pointed out that drug formulations used in dental treatments typically have a low surface tension so that the drug formulation penetrates through the tiny cavities to the site of the wound.

2.5.2 Contact Angle

The contact angle between a liquid and a solid is the angle within the body of the liquid formed at the gas–liquid–solid interface. This is geometrically determined by drawing a tangent from the contact point along the gas–liquid interface, as shown in Figure 2.5.

If the contact angle between a liquid and a solid is <90°, the liquid will wet the surface and spread over it. If the contact angle is ≥90°, the liquid will stay on the surface as a bead. Therefore, the contact angle between a liquid and a solid is dependent on the nature of the liquid as well as the surface characteristics of the solid. Both of these factors are critical when formulating nasal or ophthalmic drop dosages. The drop size and drop weight delivered from the bottles are dependent on the geometry of the orifice and the surface characteristics of the material used for the primary container, as well as the surface tension of the formulation. Van Santvliet et al. [28–30] have studied the influence of the formulation surface tension and the dropper tip angle for ophthalmic dosages. Their data suggested that the drop weight, which equates the dose weight, is dependent on the dropper angle, capillary orifice, and surface tension of the formulation. The material used for making the dropper (glass vs plastic) is also suspected to play a role in the dose weight of liquid dosages to be delivered via capillary droppers.

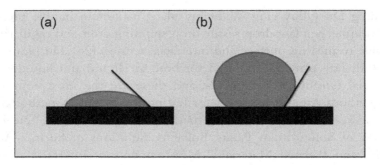

Figure 2.5 Contact angle of a water drop with a solid surface; <90° indicates wetting (a) and ≥90° indicates nonwetting (b).

2.6 CONCLUSION

Surfactants are ubiquitous in nature. Both natural and synthetic surfactants are used in pharmaceutical formulations. The FDA-IIG database lists numerous surfactants, indicating that surfactants are important excipients in all dosage forms. Surface phenomena including surface tension and contact angle studies play a critical role in formulation development, as well as in the design or selection of container closure systems for liquid dosages.

REFERENCES

[1] Fuhrhop JH, Koning J. In: Stoddart JF, editor. Membranes and molecular assemblies: the synkinetic approach. London: Royal Society of Chemistry; 1994.
[2] Fuhrhop J-H, Wang T. Bolaamphiphiles. Chem Rev 2004;104:2901–37.
[3] Ikeda S, Ozeki S, Tsunoda M-A. Micelle molecular weight of dodecyldimethylammonium chloride in aqueous solutions, and the transition of micelle shape in concentrated NaCl solutions. J Colloid Interface Sci 1980;73(1):27–37.
[4] Jones MN, Chapman D. Micelles, monolayers, and biomembranes. New York: Wiley-Liss; 1995. p. 68.
[5] Sidim T, Acar G. Alcohols effect on critic micelle concentration of polysorbate 20 and cetyl trimethyl ammonium bromine mixed solutions. J Surfactants Deterg 2013;16:601–7.
[6] Maillette de Buy Wenniger L, Beuers U. Bile salts and cholestasis. Dig Liver Dis 2010;42(6):409–18.
[7] Small DM. Size and structure of bile salt micelles: influence of structure, concentration, counterion concentration, pH, and temperature. Adv Chem 1968;84:31–52.
[8] Griese M. Pulmonary surfactant in health and human lung diseases: state of the art. Eur Respir J 1999;13(6):1455–76.
[9] Jones MN, Chapman D. Micelles, monolayers, and biomembranes. New York: Wiley-Liss; 1995. p. 8–10.
[10] Porter CJ, Trevaskis NL, Charman WN. Lipids and lipid-based formulations: optimizing the oral delivery of lipophilic drugs. Nat Rev Drug Discov 2007;6:231–48.
[11] Downing DT, Strauss JS. Variability in the chemical composition of human skin surface lipids. J Inv Dermatol 1969;53(6):322–7.

[12] Pappas A. Epidermal surface lipids. Dermatoendocrinol 2009;1(2):72–6.
[13] Pople PV, Singh KK. Development and evaluation of topical formulation containing solid lipid nanoparticles of vitamin A. AAPS PharmSciTech 2006;7(4):91.
[14] Kulkarni CV, Moinuddin Z, Patil-Sen Y, Littlefield R, Hood M. Lipid-hydrogel films for sustained drug release. Int J Pharm 2015;479(2):416–21.
[15] Cannon JB. Drug product development process for lipid-based drug delivery systems. Am Pharm Rev October 2012;15(6).
[16] Kalepu S, Manthina M, Padavala V. Oral lipid-baseddrugdeliverysystems – an overview. Acta Pharm Sin B 2013;3(6):361–72.
[17] Müllertz A, Ogbonna A, Ren S, Rades T. New perspectives on lipid and surfactant based drug delivery systems for oral delivery of poorly soluble drugs. J Pharm Pharmacol 2010;62(11):1622–36.
[18] Bos JD, Meinard MM. The 500 Dalton rule for the skin penetration of chemical compounds and drugs. Exp Dermatol 2000;9:165–9.
[19] Karande P, Jain A, Ergun K, Kispersky V, Samir Mitragotri S. Design principles of chemical penetration enhancers for transdermal drug delivery. Proc Natl Acad Sci USA 2005;102(13):4688–93.
[20] Adamson AW. Physical chemistry of surfaces. New York: John Wiley and Sons Inc; 1990. p. 537–9.
[21] Barnes GT, Gentle IR. Interfacial science an introduction. New York: Oxford University Press; 2005. p. 133.
[22] Kulkarni V, Shaw C. Formulations and characterization of nasal spray. Inhal Mag June 2012.
[23] Estroff LA, Hamilton AD. Water gelation by small organic molecules. Chem Rev 2004;104:1201–17.
[24] Raghavan SR. Distinct character of surfactant gels: a smooth progression from micelles to fibrillar networks. Langmuir 2009;25(15):8382–5.
[25] Patel HK, Rowe RC, McMahon J, Stewart RF. Properties of cetrimide/cetostearyl alcohol ternary gels; preparation effects. Int J Pharm 1985;25(2):237–42.
[26] Jouyban A, Fathi-Azarbayjani A. Experimental and computational methods pertaining to surface tension of pharmaceuticals. In: Toxicity and drug testing. In Tech; 2012. p. 47–70.
[27] Pecora JD, Guimaraes LF, Savioli RN. Surface tension of several drugs used in endodontics. Braz Dent J 1991;2:123–7.
[28] Sklubalová Z, Zatloukal Z. Study of eye drops dispensing and dose variability by using plastic dropper tips. Drug Dev Ind Pharm 2006;32(2):197–205.
[29] Van Santvliet L, Ludwig A. Determinants of eye drop size. Surv Ophthalmol 2004;49(2):197–213.
[30] Van Santvliet L, Ludwig A. The influence of penetration enhancers on the volume instilled of eye drops. Eur J Pharm Biopharm 1998;45(2):189–98.
[31] Kulkarni V. Liposomes in Personal Care Products. Delivery system handbook for personal care and cosmetic products. In: MR Rosen, editor. William Andrew Publishing, 2005, pp.285-302.

CHAPTER 3

Drug Delivery Vehicles

Contents

3.1 INTRODUCTION

As discussed in the introductory chapter, drug substances cannot usually be administered or applied directly to the body due to safety reasons. As a result, drug substances need to be either dissolved or suspended in a suitable matrix so that they can be safely injected, swallowed, or applied to the skin (either intact or compromised) or mucosal membranes. The matrix in which the drug substance is dissolved or suspended is considered a "vehicle" for carrying the drug to its intended target. If the drug substance dissolves in the formulation, the solvent (or solvent system) is the drug delivery vehicle. However, not all drugs can be delivered in the form of a solution because a large number of drug substances are poorly soluble in either water or water/cosolvent mixtures. In this case, the drug substance is suspended in a "structured delivery vehicle" (either solid–in–liquid or liquid–in–liquid). The most common structured-delivery vehicles are emulsions, gels, liposomes, polymers, dendrimers, solid–lipid nanoparticles, and composite nanoparticles and have been used for injectable/infusible liquids, subcutaneous gel injections, semi–solid topical products (e.g. gels, creams, lotions, ointments), and oral suspensions. An advantage of structured drug delivery systems is that the drug (whether hydrophilic or hydrophobic in nature) can be entrapped in the formulation matrix, and that this encapsulated drug can be formulated into a stable delivery vehicle such that the drug can be safely delivered to its target destination. Compared to systemic delivery, delivering the drug to the target site enables the dose of the drug to be lowered, thus reducing possible toxicity and side effects. Structured drug delivery systems can also be engineered for sustained drug release, drug release triggered by

Essential Chemistry for Formulators of Semisolid and Liquid Dosages
http://dx.doi.org/10.1016/B978-0-12-801024-2.00003-0

21

external stimulus, and providing protection to the drug substance from degradation—thereby extending the shelf life of the drug product compared to unprotected formulations (e.g. simple solutions).

3.2 EMULSION DRUG DELIVERY SYSTEMS

Emulsions consist of two immiscible phases (oil and water). One of the phases is dispersed as droplets within the other phase, and the formulation is stabilized using a combination of surfactants, emulsifiers, and thickeners. There are two main categories of emulsions: macroemulsions and microemulsions. Macroemulsions are thermodynamically unstable and are kinetically stabilized using elevated temperature and high shear during the mixing process. Microemulsions are formed spontaneously in the presence of the appropriate concentration of certain surfactants and cosolvents. Oil-in-water macroemulsions that contain very small internal phase droplets (generally <500 nm) are sometimes called nanoemulsions. Suitable surfactants can be selected by matching the "required HLB (Hydrophile Lipophile Balance) number" of the oil to be emulsified with the HLB of two or more surfactants [1]. The Active Pharmaceutical Ingredient (API) is either dissolved in the oil or water phase, or it may be suspended in the formulation. An existing marketed product for intravenous infusion that is based on an emulsion-delivery system is Intralipid®, which is a nanoemulsion consisting of 10%, 20%, or 30% w/v soybean oil in water, and is stabilized with various phospholipids. It is practically isotonic with blood, has a pH of approximately 8, and a particle size of 200 – 500 nm.

Microemulsions have been described as optically isotropic/transparent and thermodynamically stable liquid solutions [2] consisting of water, oil, and surfactant. A generalized ternary phase diagram for the formation of microemulsions is shown in Figure 3.1. In many cases, cosurfactants and/or cosolvents are needed to generate microemulsions. The particle size of microemulsions generally ranges from 5 to 50 nm. Microemulsions are sometimes called "micellar emulsions" [3]. There are four types of microemulsions:

Type I: the surfactant is dominantly soluble in water, resulting in an oil-in-water (o/w) microemulsion.

Type II: the surfactant is mainly soluble in the oil phase, resulting in water-in-oil (w/o) microemulsion.

Type III: a "surfactant-rich" middle phase coexists with larger water and oil "surfactant-poor" phases, resulting in a three-phase system.

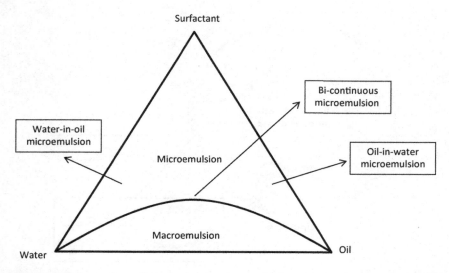

Figure 3.1 Ternary phase diagram showing microemulsion and macroemulsion regions (see Ref. [2]).

Table 3.1 Comparison of macroemulsions and microemulsions

Macroemulsion	Microemulsion
Thermodynamically unstable	Thermodynamically stable
Kinetically stabilized	Formed instantly
Appearance: Opaque, milky liquid	Appearance: Translucent to clear liquid
Particle size typically >0.1 µm (range 0.1–100 µm)	Particle size typically <0.1 µm (range 0.01–0.1 µm)
Shelf life could be up to 3 years	Stable for very long time
Low surfactant/emulsifier levels	High levels of surfactant, co-surfactants, or cosolvents

Type IV: a single-phase (isotropic) micellar solution, which forms upon addition of a sufficient quantity of surfactant and cosolvent.

For comparison between macro- and microemulsions, see Table 3.1.

3.3 LIPOSOME DRUG DELIVERY SYSTEMS

Lipids form spherical vesicles (known as liposomes) consisting of an aqueous core and a lipophilic bilayer enclosing one or more aqueous compartments. In solution, both the inner core of the liposomes and the continuous phase are aqueous, and the two aqueous components are separated by a

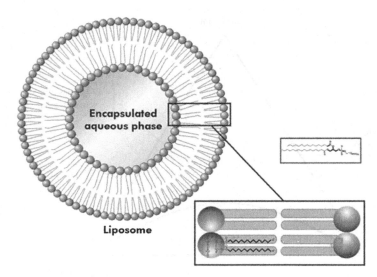

Figure 3.2 Schematic representation of a liposome. Aqueous phase is encapsulated by a bilayer of phospholipids. The chemical structure of the phospholipid is shown and superimposed on the schematic lipid representation.

lipid bilayer. The liposome structure is shown schematically in Figure 3.2. Water-soluble drug substances can be encapsulated in the aqueous core of the liposomes, and lipophilic drugs can be entrapped in the bilayer. This enables liposomes to deliver both hydrophilic and lipophilic drug substances to the intended site of action. Liposomes comprise phospholipids, which are also the main building blocks of cell membranes—thereby improving the biocompatibility of the drug delivery system. Recent advances in the liposome technology have focused on "targeted delivery". These targeted-delivery liposomes (also known as "immuno-liposomes") have specific binding molecules on their outer surface that bind to the target cells and deliver their "cargo." Liposomes that are capable of circulating in the blood for extended periods are engineered by modulating the lipid composition, size, charge, and surface structure. For example, the circulation times of liposomes prepared by using lipids modified with polyethylene glycol (PEG), called PEGylated lipids, are considerably enhanced. Such liposomes are known as "stealth liposomes" because they are not detected by the mononuclear phagocyte system [4]. Liposomes made with mainly cationic lipids result in a positive charge on the liposomes, and are called "cationic liposomes." Cationic liposomes have been used for DNA delivery.

Liposomes are categorized by their lamellarity and size:
- multivesicular liposomes have several smaller vesicles contained within a large vesicle
- multilamellar vesicles (MLVs) have multiple layers of concentric bilayers separated by layers of the aqueous phase
- large unilamellar vesicles (LUVs)
- small unilamellar vesicles (SUVs)

3.3.1 Methods of Making Liposomes

There are four common processes for making liposomes:
1. Solvent evaporation: Lipids are dissolved in a suitable organic solvent, such as chloroform. The solution is transferred into a round-bottomed flask, and the solvent is evaporated using a rotary evaporator. This forms a thin film of lipid on the inner wall of the flask. The lipid film is then hydrated with water (or appropriate buffer solution), and a brief sonication allows the lipid film to be dispersed into the aqueous phase. This aqueous phase consists of large multilamellar liposomes, which can be extruded through a membrane of suitable pore size (100 nm) to obtain unilamellar liposomes. The suspension of liposomes may be subjected to probe sonication to form small unilamellar vesicles that are then harvested by ultracentrifugation.
2. Ethanol injection: Lipids are dissolved in ethyl alcohol. The resulting solution is quickly injected through a small-bore needle into a fast-mixing aqueous phase (water or buffer), thereby producing tiny droplets of the lipid within the aqueous phase. As soon as the alcohol is exposed to water, it dissolves allowing the lipid to quickly hydrate and self-assemble into large multilamellar vesicles. This dispersion can then be extruded through a 100 nm membrane, or subjected to probe sonication, to make small unilamellar vesicles.
3. Detergent dialysis: In the presence of certain surfactants, lipids can be dissolved in the aqueous phase. The surfactants are subsequently removed from the aqueous phase by dialysis, leaving the well-hydrated lipid to form liposomes. The liposome dispersion can then be extruded or subjected to probe sonication.
4. High-pressure homogenization: A mixture of lipid and aqueous phases is homogenized at high speeds (10,000–14,000 rpm). The raw dispersion is immediately passed through a high-pressure homogenizer, such as a Microfluidizer®, in which the product is subjected to very high pressures (18,000–30,000 psi, or higher). The resulting product is a unilamellar

liposome of uniform diameter. High-pressure homogenization is suitable for making large-sized batches of 500 L or more.

Two methods are used to encapsulate drug substances into liposomes: passive loading and active (assisted) loading. In passive loading, the lipid and drug substance ("cargo") are mixed together in the water/buffer phase. During the formation of the liposomes, the drug gets encapsulated within the vesicle. The nonencapsulated drug can be dialyzed out, leaving only the encapsulated drug in the delivery system. Freeze and thaw (FAT) techniques, and creating pH or salt gradients between the encapsulated and external aqueous phase are some of the active (assisted) liposome-loading techniques. Creating a pH gradient between the encapsulated aqueous compartment and the external aqueous phase of 3 or more pH units (e.g., pH 3.5 inside of the liposomes and pH 7.5 outside of the bilayers) has been shown to transfer amine-like active ingredients from the external aqueous phase to the inside of the liposome [5].

The FDA has published guidance on Liposome Drug Products (http://www.fda.gov). This guidance recommends characterizing the liposome drug product for attributes such as:

- Morphology, including lamellarity determination
- Net charge
- Entrapment volume
- Particle size
- Phase-transition temperature
- In vitro release of the drug substance from the liposome drug product
- Osmotic properties
- Light-scattering index

The integrity and size of the liposomes, as well as the stability of the lipids, are among the parameters that should be tested as part of a stability program for liposome drug products. The stability of both the unloaded liposome product (placebo) and the active liposome product should be evaluated.

Some of the existing drug products based on liposome drug delivery systems are shown in Table 3.2.

3.4 OTHER NANODRUG DELIVERY SYSTEMS

- Polymeric nanoparticles: Biodegradable nanoparticles of poly DL-lactic-co-glycolic acid (PLGA), polycaprolactone, and PEG copolymers have been used for drug delivery. Emulsification, followed by solvent evaporation by freeze drying, is one of the common methods of preparing

Table 3.2 Examples of drug products based on the liposome drug delivery system

Trade name	Active pharmaceutical ingredient (API)
Abelcet	Amphotericin B lipid complex injection
AmBisome	Amphotericin B liposome for injection
Daunoxome	Daunorubicin liposome injection
Definity	Perflutren lipid microsphere injectable suspension
DepoCyte	Cytarabine liposome injection
DepoDur	Morphine sulfate extended-release liposome injection (epidural injection)
Doxil	Doxorubicin hydrochloride liposome injection
Myocet	Liposome-encapsulated doxorubicin-citrate complex
Visudyne	Verteporfin for injection, reconstituted in liposome

nanoparticles of drug-loaded PLGA. Studies on the use of tobramycin-loaded PLGA nanoparticles for pulmonary drug delivery, and the delivery of vaccines to the lungs via nanoparticle encapsulation have been reported [6,7]. Compared to liposomes, polymeric nanoparticles are rigid, more stable, and can be engineered/tailored for specific needs. Technological advances have enabled the controlled formation of nanoparticles, and production on a large scale is becoming feasible. As a result, there has been considerable interest in the literature related to development of anticancer drugs with nanoparticle carriers. Single or multiple drugs can be loaded into polymeric nanoparticles and delivered with reduced side effects [8,9].

- Dendrimers: Dendrimers have a globular, monodispersed nanometric polymeric architecture in which the atoms are arranged in many branches along a central backbone of carbon atoms. Dendrimers have a precise molecular weight, shape, and size, and are emerging as drug delivery vehicles [8,10,11].
- Other nanodrug-delivery systems include metal colloids, nanocrystals (quantum dots), fullerenes, and carbon nanotubes. Although most of these nanodrug delivery systems are still in the exploratory stages for pharmaceutical applications, in the near future they are likely to become more widespread in the marketplace. In fact, FDA-approved sunscreens using nanoscale TiO_2 and ZnO for UV light protection already exist.

Particle size, morphology, surface area, surface-charge zeta potential, and drug-loading capacity are critical properties for characterizing nano-drug-delivery systems. As discussed in the chapters on particle size and microscopy, electron microscopy is a very useful tool for nanoparticle characterization.

3.5 CONCLUSION

Structured drug delivery systems such as emulsions, nanoemulsions, and liposomes have been successfully utilized in drug products. Other nanoparticle systems are under development, and are showing great potential as future pharmaceutical delivery platforms.

REFERENCES

[1] Ramchandani M, Toddywala R. Formulation of topical drug delivery systems. In: Ghosh TK, Pfister WR, Yum SI, editors. Transdermal and topical drug delivery systems. Buffalo Grove, IL: Interpharm Press Inc.; 1997. p. 539–78.

[2] Gupta S, Moulik SP. Biocompatible microemulsions and their prospective use in drug delivery. J Pharm Sci 2008;97:22–45.

[3] Adamson AW. Physical chemistry of surfaces. New York: John Wiley and Sons, Inc.; 1990. p. 539–44.

[4] Nag OK, Awasthi V. Surface engineering of liposomes for stealth behavior. Pharmaceutics 2013;5:542–69.

[5] Waterhouse DN, Mayer LD, Cullies PR, Madden TD, Bally MB. The liposomal formulation of doxorubicin. Method Enzymol 2005;391:71–97.

[6] Ungaro F, d'Angelo I, Coletta C, d'Emmanuele di Villa Bianca R, Sorrentino R, Perfetto B, et al. Dry powders based on PLGA nanoparticles for pulmonary delivery of antibiotics: modulation of encapsulation efficiency, release rate and lung deposition pattern by hydrophilic polymers. J Control Release 2012;157(1):149–59.

[7] Garcia Contreras L, Awashthi S, Hanif SNM, Hickey AJ. Inhaled vaccines for the prevention of tuberculosis. J Mycobac Dis 2012;S1:1–13.

[8] Hu C-MJ, Aryal S, Zhang L. Nanoparticle-assisted combination therapies for effective cancer treatment. Ther Delivery 2010;1(2):323–34.

[9] Beck_Broichsitter M, Gauss J, Gessker T, Seeger W, Kisssel T, Schmehl T. Pulmonary targeting with biodegradable salbutamol loaded nanoparticles. J Aerosol Med Pulm Drug Deliv 2010;23(1):47–57.

[10] Ravi Kumar MNV, Muzzarelli RAA, Muzzarelli C, Sashiwa H, Domb AJ. Chitosan chemistry and pharmaceutical perspective. Chem Rev 2004;104:6017–84.

[11] Tripathi S, Das MK. Dendrimers and their applications as novel drug delivery carriers. J Appl Pharm Sci 2013;3(9):142–9.

CHAPTER 4

Formulating Creams, Gels, Lotions, and Suspensions

Contents

4.1 INTRODUCTION

Creams, lotions, gels, ointments, and pastes are examples of different types of formulations that are used as semisolid dosages. Semisolid dosages are mostly administered topically, transdermal, or (in the case of some gels) via subcutaneous injection and can contain many ingredients (5–10 or more). Solution formulations, on the other hand, are generally simpler and contain fewer ingredients. As a result, semisolid dosages are more complex than solutions and are more challenging to make. A decision tree for the classification of topical dosage form nomenclature has been published by Buhse et al. [1] and is shown in Figure 4.1.

4.2 CREAMS AND LOTIONS

Creams and lotions are examples of emulsions—thermodynamically unstable two-phase systems consisting of at least two immiscible liquids (e.g., oil and water), one of which is dispersed in the form of small droplets throughout the other. The phase present in the form of droplets is known as the internal (or dispersed) phase, and the matrix in which the droplets are dispersed is known as the external (or continuous) phase. Surfactants (emulsifiers) are added to stabilize the system by lowering the surface tension (interfacial tension) between the two phases. In an emulsion, the drug substance may be contained in either the oil phase or the water phase depending on the solubility of the drug. Emulsion systems may enhance the activity of the drug by acting as

Essential Chemistry for Formulators of Semisolid and Liquid Dosages
http://dx.doi.org/10.1016/B978-0-12-801024-2.00004-2
29

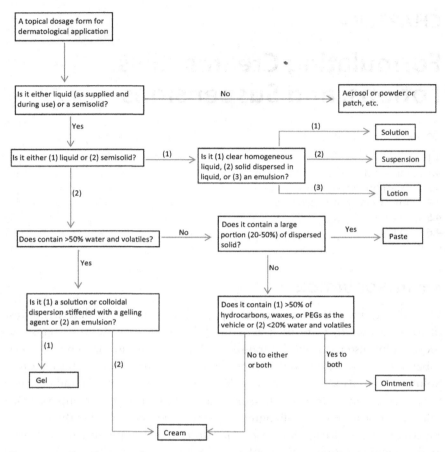

Figure 4.1 Classification of the semisolid dosage form shown here reflects the current thinking of US Food and Drug Administration. *From Ref. [1] (with permission from publisher).*

either a solvent for the API (Active Pharmaceutical Ingredient) or a penetration enhancer. Creams typically contain >20% water and "volatiles" (volatiles are measured by heating at 105 °C until the constant weight is achieved) and <50% of hydrocarbons, waxes, or polyethylene glycols. Creams come in the form of either oil-in-water emulsions (in which the oil phase is dispersed as droplets in the water phase; "O/W") or water-in-oil emulsions (in which the water phase is dispersed as droplets in the oil phase; "W/O"). Lotions, on the other hand, tend to be more liquid-like than creams and have a lower viscosity. Most of the existing lotions in the marketplace are O/W emulsions. The terms "cream" and "lotion" are included under the general category of "emulsion" in the US Pharmacopeial Convention (USP); however, it is reported that the term "lotion" is likely to be phased out [2].

A few examples of pharmaceutical creams and their ingredients are listed below.

- Antivirals, for example, Denavir® (for cold sores): 1% penciclovir; inactive ingredients include cetostearyl alcohol, mineral oil, polyoxyl 20 cetostearyl ether, propylene glycol, purified water, and white petrolatum.
- Antihistamine creams, for example, Benadryl® itch cream; 2% diphenhydramine hydrochloride and 0.1% zinc acetate; inactive ingredients include cetyl alcohol, diazolidinyl urea, methylparaben, polyethylene glycol monostearate 1000, propylene glycol, propylparaben, and purified water.
- Antifungal creams, for example, Ketoconazole cream; 2% ketoconazole, inactive ingredients include propylene glycol, purified water, cetyl alcohol, stearyl alcohol, isopropyl myristate, sorbitan monostearate, polysorbate 60, polysorbate 80, and sodium sulfite.
- Anesthetic creams, for example, PLIAGLIS® Cream; 7% each of lidocaine and tetracaine; inactive ingredients include dibasic calcium phosphate, methylparaben, petrolatum, polyvinyl alcohol, propylparaben, purified water, and sorbitan monopalmitate.

The inactive ingredients listed in the above-mentioned examples show that creams comprise a complex mixture of surfactants, viscosity builders, emulsion stabilizers, preservatives, antioxidants, pH-adjusting agents, and chelating agents. When creams and lotions are formulated, the inactive ingredients/functional agents are chosen to provide the desired product viscosity (so that they can be easily applied on the skin) and to ensure product stability (to achieve an acceptable shelf life).

In simple terms, a generalized procedure for making an emulsion would be:

1. In Tank #1: combine the oil-soluble ingredients (e.g., mineral oil, vegetable oil, and/or mixtures of fatty acid, fatty alcohols, and fatty esters). While mixing, heat the oil phase to a temperature above the melting point of the highest-melting component (generally 70 °C or higher)
2. In Tank #2: mix the water-soluble ingredients in water. Heat the water phase to the same temperature as the oil phase
3. For an O/W emulsion, add the oil phase to the water phase and mix vigorously (use high-shear mixing if required) to form a homogeneous dispersion of small oil droplets within the water phase (in the case of W/O emulsions, the water phase is added to the oil phase)
4. Continue mixing (using lower shear) and cool the system to room temperature. At a given temperature, the emulsion will form (usually accompanied by a change in the appearance of the mixture and an increase in the viscosity)

5. If the drug substance has good thermal stability (i.e., stable at the elevated mixing temperature), it may be added to the phase in which it is soluble at the beginning of the process. If thermal stability is an issue, it may be added after the emulsion is formed. In this case, a suitable dispersing process (such as a mill) may be required to make a uniform product

Thermodynamically, macroemulsions (creams and lotions in which the dispersed-phase droplet size is more than 0.1 µm) are unstable because the Gibbs free energy of the system is positive. Imagine a system in which an oil phase, represented by a single large drop (see Figure 4.2) of surface area A1, is immersed in an aqueous phase. After the emulsion is formed, this single drop subdivides into several smaller droplets with total area surface A2 (A2≫A1; see Table 4.1). It is assumed that the interfacial tension between the oil and the water is the same for the single large drop (preemulsion) and the small droplets (postemulsion), provided that the droplets are larger than

Figure 4.2 Schematic representation of an oil phase in an aqueous phase before and after emulsification. Imagine the oil phase as a single drop in a water phase that is divided into several small droplets after emulsification. Calculations in Table 4.1 show that as the droplet size decreases the number of droplets and the combined surface area increase enormously.

Table 4.1 As the droplet size of oil phase is reduced, the number of droplets and combined surface area of all droplets increase enormously when forming an oil-in-water emulsion

Droplet size, µm	Number of drops	Surface area, cm²
26,730 (~2.7 cm)	1	22.4
20	2.4×10^9	30,000
10	1.9×10^{10}	60,000
1	1.9×10^{13}	600,000
0.5	1.5×10^{14}	1,200,000

In the above-mentioned calculations it is considered that 10 mL of oil phase is dispersed in the water phase, assuming that initially all of the oil phase (V = 10 mL) is in the form of a single drop (see Figure 4.2).

0.01 μm [3]. Two factors contribute to the change in free energy in going from state I (preemulsion) to state II (postemulsion): (1) A surface-energy term (which is positive) that is equal to $\Delta A \gamma_{12}$ (in which $\Delta A = A_2 - A_1$, and γ_{12} is the interfacial tension between the two phases); (2) An entropy-of-dispersion term (which is also positive, because producing a large number of droplets is accompanied by an increase in configurational entropy, ΔS^{conf}). From the second law of thermodynamics:

$$\Delta G^F = \Delta A \gamma_{12} - T \Delta S^{conf}$$

In most cases, $\Delta A \gamma_{12} \gg T \Delta S^{conf}$ indicating that ΔG^F is positive. Evidence of this can be seen by the fact that formation of the emulsion is nonspontaneous; thus, macroemulsions are thermodynamically unstable. Consequently, in the absence of a stabilization mechanism the emulsions will break down.

One method by which emulsions break down is coalescence of the dispersed-phase droplets to form larger droplets. Over time, these larger droplets separate from the continuous phase—either floating on the top or sinking to the bottom, depending on the size of the droplets and the relative density of the droplets and the continuous phase. This type of phase separation is known as "creaming". From Stoke's equation, the rate of coalescence (rate of creaming) can be estimated.

Stoke's equation:

$$\upsilon = d^2 (\rho - \rho_0)*g/18\eta$$

in which υ, creaming rate; d, diameter of the droplet; ρ, density of the droplet; ρ_0, density of the dispersion medium; g, acceleration due to gravity; and η, viscosity of the dispersion medium.

For given oil and water phases, ρ, ρ_0, and g are fixed. Therefore, for an emulsion, the rate of coalescence decreases as the diameter of the droplets decreases and the viscosity of the dispersion medium increases. In other words, emulsions can be kinetically stabilized if the droplets are made sufficiently small, and the viscosity of the dispersion medium is sufficiently high.

To prevent coalescence, stabilizers such as surfactants, emulsifiers, thickeners, and polymers are added to the system. These additives form an energy barrier between the droplets and make the system kinetically stable for a finite period. Understanding and controlling the factors that influence the formation and stability of the emulsion are critical to producing a stable product with an adequate shelf life (three years is typically expected for pharmaceutical creams or lotions). For testing the stability of emulsions, particle/droplet size distribution, viscosity, and rheology are key physical parameters.

4.3 GELS, OINTMENTS, AND SUSPENSIONS

A gel consists of suspended particles in a dispersion medium. These particles (or gelling agents) undergo a high degree of cross-linking or association when hydrated, forming an interlaced three-dimensional structure that provides stiffness to a solution or dispersion. Gels tend to be shear-thinning (spread easily with applied pressure or friction) and are typically used for dosages that are applied externally to the skin [1]. Gels can be water- or water/alcohol-based systems. Gelling agents include aluminum hydroxide, bentonite, aluminum or zinc soaps, magnesium aluminum silicates, xanthan gum, colloidal silica, starch, cellulose derivatives, and carbomers (polyacrylates). Typical inactive ingredients include preservatives, antioxidants, pH-adjusting agents, and chelating agents.

A generalized procedure for making a gel would be:
1. Mix the inactive ingredients in the solution base (water or water/alcohol)—typically at room temperature. Heating might be needed, however, to dissolve some of the ingredients
2. While continuing to mix, add the active ingredient
3. Finally, add the gelling agent to increase the viscosity of the mixture and trap/suspend any solid particulate material. Reduce the mixing shear as the gel viscosity increases

Ointments are viscous, semisolid preparations containing either dissolved or suspended ingredients. Although proprietary ointment bases exist, the most common ointment bases comprise a mixture of petrolatum and mineral oil or petrolatum and waxy/fatty alcohol combinations. The ratio and grades of the ointment base components are selected to give the desired finished product viscosity/spreadability. A typical procedure for making an ointment would be:
1. Add the ointment base ingredients to the mixing vessel and heat above their melting temperature. Mix the ointment base at low mixing speeds and low shear until it is molten
2. Add the other ingredients and increase the mixing speed to form a uniform dispersion (external powder eductors can be used to incorporate solid ingredients)
3. Finally, cool the mixture to room temperature, reducing the mixing speed as the viscosity of the ointment increases

A suspension is a two-phase system consisting of finely divided solid material dispersed in a liquid. Suspending agents are used to reduce the rate of sedimentation of the solid material by forming films around the suspended particles (thereby decreasing interparticle attraction) and by acting as thickening agents (increasing the viscosity of the solution). Typical suspending agents include some natural, semisynthetic, and synthetic hydro-colloids and clays (see Chapter 5). The procedure for making suspension formulations is similar to that for making gels.

4.4 FORMULATING WITH LIPOSOMES

Liposomes have an aqueous solution core surrounded by a hydrophobic membrane, in the form of a lipid bilayer. During the formulation process, hydrophilic materials can be dissolved in the aqueous core and hydrophobic materials made to associate with the bilayer. Consequently, liposomes can be used in formulations as "carriers" for both hydrophilic and hydrophobic ingredients. The cell membranes in human skin are also made up of lipid bilayer structures. The lipid bilayers of the liposomes in formulated products can "fuse" with the bilayers in the cell membrane, thereby delivering the liposome contents to a site of action. In this way, liposomes can be used as a vehicle for the administration of nutrients and pharmaceutical drugs.

The presence of surfactants in a formulation will compromise the integrity of liposomes. As a result, stability becomes a major concern when attempting to formulate liposomes in emulsion systems. For topical or transdermal routes of administration, liposomes can be safely formulated in gel-type dosages. Gels made from carbomer, cellulose derivatives, and hyaluronic acids are most suit-able for formulating with liposomes. Entrapping the liposomes in a thickened matrix helps to prolong their shelf life by reducing the possibility of liposome–liposome collisions. For injectable liposome formulations, the liposomes can be freeze-dried and supplied with a separate suitable carrier medium (such as saline- or dextrose-injection solutions). The liposomes are resuspended by mix-ing with the carrier medium immediately prior to injecting into the body.

Some of the critical factors that affect the stability of liposomes include [4]:
- pH of the formulation: A finished-formulation pH of 6.5 is ideal, because, at this pH, the rate of lipid hydrolysis is lowest.
- Storage temperature: Liposomes are very susceptible to temperatures that promote oxidation and leakage of the entrapped cargo. Therefore, storage at 2-8 °C is ideal. Additionally, it is critical not to subject loaded liposomes to freeze and thaw conditions as it is known that the loaded cargo is likely to leak after freeze–thaw stress.
- Container-closure system: The selection of the container-closure system used for storing liposome formulations is crucial. Liposomes are not com-patible with certain plastic materials. For injectable liposome suspensions, testing compatibility with the elastomeric stoppers to be used with the

injection vials is essential. Using glass ampoules rather than stoppered injection vials is often safer. As lipids are susceptible to photooxidation, protecting them from light during storage is highly recommended.

- Infusion tubing: For infusible liposome formulations, establishing compatibility of the liposome suspensions with intravenous tubing is critical, because this tubing is made of synthetic plastic materials. The product label needs to specify the parts/types of tubing that can be used during drug administration.

4.5 EXCIPIENTS

Categories of excipients generally used in semisolid and liquid dosages are discussed in USP <1059>. Some examples are presented below.

- Preservatives: All nonsterile dosage forms need to be protected against microbial growth. This is commonly accomplished by the addition of preservatives to the formulation. Several factors are involved in the selection of a preservative, including solubility in the water phase, the partition coefficient between the oil and water phases, taste/odor (for oral formulations), the type of formulation (solution, emulsion, spray, ointment), pH of the product, compatibility with the drug substance and other excipients, possible side effects, and the target patient population (infant/pediatric, adult, geriatric) [5]. The relationship between pH and antimicrobial activity is often complex—for example, benzoic acid is both antifungal and antimicrobial; however, antifungal activity is less susceptible to pH than antimicrobial activity. Quaternary ammonium compounds are incompatible with anionic surfactants, anionic excipients, or gels composed of carbomer (anionic polymer). An ideal preservative exhibits a wide spectrum of antimicrobial activity, remains active throughout product manufacture to the end of the product shelf life, does not compromise the quality and performance of the product or package, and does not adversely affect patient safety. Regulatory agencies generally require justification for inclusion, proof of efficacy, and safety information not only on preservatives, but also for most of the excipients used in formulations. The required concentration of the preservative is determined by separate studies on "minimum inhibitory concentration", that is, the lowest concentration of an antimicrobial preservative that will inhibit the visible growth of a microorganism after overnight incubation. Table 4.2 lists some of the

Table 4.2 List of antimicrobial preservatives used in pharmaceutical formulations compiled from the FDA inactive ingredients guide

General category	Preservatives	Dosage form
Parabens	Butylparaben	Oral, suspensions, topical
	Ethylparaben	Topical
	Methylparaben	Oral, topical
	Propylparaben	Oral, topical
	Propylparaben sodium	Oral
	Sodium ethylparaben	Oral capsule, soft gelatin
Alcohols	Benzyl alcohol	Topical, creams, lotions, gels, nasal sprays, intravenous (IV)/intramuscular (IM)/ subcutaneous (SC)-injection solutions
	Propylene glycol	Ophthalmic, oral, topical
	Chlorobutanol	Injections, nasal sprays, ophthalmic, topical
	Phenol	IV (infusion), IM, IV, SC
	3-Cresol (meta cresol)	IV (infusion), IM, IV, SC
	4-Chlorocresol	Topical
	4-Chloroxylenol	Topical
	Phenylethyl alcohol	Nasal, ophthalmic
	Phenylmercuric acetate	Topical, ophthalmic, IM, nasal
	Phenylmercuric nitrate	Ophthalmic, IM
	Thimerosal	Topical, ophthalmic, SC, IM, IV
	Phenoxyethanol	Topical
Acids/salts	Sorbic acid	Topical, ophthalmic, oral
	Potassium sorbate	Topical, ophthalmic, oral
	Benzoic acid	Topical, oral, IV, IM
	Sodium benzoate	Topical, oral, IV, IM, dental
	Boric acid	Ophthalmic, topical, IV
	Sodium propionate	Oral
Quaternary ammonium compounds (QACs)	Benzalkonium chloride	Ophthalmic, nasal, inhalation, IM, topical
	Cetylpyridinium chloride	Oral, inhalation, transdermal
	Benzethonium chloride	Ophthalmic, nasal, IV (infusion), injection
	Cetrimonium chloride	Topical
Formaldehyde donor	Imidurea	Topical

preservatives and their reported use in different dosage forms (compiled from the FDA Inactive Ingredient Guide [IIG]).

- Antioxidants: Drugs or excipients that are sensitive to oxygen may degrade faster due to oxidation during manufacture or product storage. Oxidation is caused by free radicals, metal ions in the formulation, heat applied during processing, or photosensitivity of the active or inactive ingredients in the formulation. Ingredients that are susceptible to oxidation include unsaturated oils, lipids, flavors, and essential oils. To delay the onset of oxidation or slow down the oxidation process, antioxidants are added to the formulation. Phenolic antioxidants block free-radical chain reactions and are most effective in protecting oils or oil-soluble ingredients against oxidative stress. Reducing-agent antioxidants are water soluble and have a lower redox potential than the active or inactive ingredients that they are protecting. They protect the formulation by sacrificially reacting with oxygen or other reactive species. As trace metal ions in the formulation can act as catalysts and promote free radical-induced oxidation, the inclusion of chelating agents can also assist in retarding oxidative degradation. Consequently, a combination of antioxidants, reducing agents, and chelating agents can be used to effectively protect a formulation from oxidation. The downside of using antioxidants in oral formulations, however, is that these excipients can be associated with gastric irritation and/or diarrhea [6]. To further protect oxygen-sensitive formulations, the manufacturing process could be performed in a controlled environment such as under a nitrogen blanket or in actinic light. Some of the antioxidants and the dosages that they are used in are listed in Table 4.3.
- Solubilizing agents: Cosolvents such as alcohol, propylene glycol, polyethylene glycols, and certain surfactants are commonly used as solubilizing agents for poorly soluble ingredients.
- Stabilizers: Macroemulsions (creams and lotions) are thermodynamically unstable systems comprising oil and water phases. These systems are kinetically stabilized using various types of polymers, gums, and gelling agents. Furthermore, the formation and stability of gels is dependent on the strength of the three-dimensional structures formed by certain polymers. Some of the thickeners and gelling agents used in FDA-approved products are shown in Table 4.4.

Table 4.3 Antioxidants used in semisolid and liquid dosages

Antioxidants	Dosage form
Ascorbyl palmitate	Oral, topical
Butylated hydroxyanisole	Oral, nasal, IV, IM, topical
Butylated hydroxytoluene	Topical, transdermal, oral, IV, IM
Cysteine	IM, SC
Potassium metabisulfite	IV, IM
Sodium metabisulfite	Dental, oral, ophthalmic, topical, inhalation
Propyl gallate	Topical, oral, IM
Sodium thiosulfate	Oral, IV, topical, ophthalmic
Vitamin E	Oral, topical
Trisodium HEDTA* (chelating agent)	Topical

*N-(2-Hydroxyethyl)ethylenediamine-N,N',N'-triacetic acid trisodium salt.

- Moisturizers and Emollients (skin protectants): These are used in topical semisolid dosages. Whereas moisturizers increase the water content of the outer layer of the skin, emollients help maintain a soft and smooth skin feel by preventing water loss by evaporation. By maintaining skin-moisture levels, moisturizers and emollients indirectly assist in the penetration of active ingredients into the skin.
- Buffering agents/pH adjusters: Acid, bases, and buffer salts are used in formulations to adjust the pH to the level needed to keep the active ingredient in the dissolved state, improve its stability, and ensure that the formula as a whole has an acceptable shelf life. Changes in the product pH from a specified range constitute a product-stability failure. Consequently, pH is one of the most commonly tested physical properties both at the time that the product is manufactured and over its shelf life.
- Penetration enhancers: In the case of some topical and all transdermal products, it is necessary that the drug substance penetrates the skin. A detailed discussion on skin penetration is presented in Chapter 2.
- Propellants: Propellants are gases at room temperature and atmospheric pressure. When compressed, however, they liquefy and provide a "reservoir of pressure" that can be used to expel products from an aerosol container. Examples of propellants include hydrofluorocarbons (e.g., HFA 134a, HFA 227) and hydrocarbons (e.g., propane, isobutane, n-butane, dimethyl ether). When the product is expelled from its container, the propellant evaporates, and the product is deposited at the site of application.

Table 4.4 Examples of gelling agents and thickeners used in pharmaceutical semisolid or liquid formulations

Ingredient category	Ingredients	Route of administration and dosages form reported in FDA-IIG
Cellulose derivatives	Carboxymethyl cellulose	Topical lotion, ophthalmic solutions, nasal spray
	Hydroxymethyl cellulose	Topical solutions
	Hydroxypropyl cellulose	Topical gel, lotion, oral suspension
	Hydroxypropylmethyl cellulose	Oral suspensions, ophthalmic solution
	Hydroxyethyl cellulose	Topical lotions, solutions, ophthalmic solutions
	Methylcellulose	Topical lotions, creams, nasal gel, ophthalmic solutions
Polymers	Carbomer	Topical gels, creams, lotions, oral suspensions
	Polycarbophil	Topical gels, solutions, suspensions
	Poloxamers	Topical gels, lotions, creams, oral suspensions
	Sodium hyaluranate	Topical gel
	Polyvinyl alcohol	Topical lotions, ophthalmic suspensions
Colloidal solids	Magnesium aluminum silicate	Topical lotion, ointments, oral liquids and suspensions
	Bentonite	Topical lotion, suppositories, topical and oral suspensions
	Microcrystalline cellulose	Topical gels, oral suspensions, nasal sprays
Natural gums	*Acacia* (gum arabic)	Oral suspensions, syrups
	Alginates	Oral suspensions
	Carrageenan	Topical lotion, oral suspensions
	Collagen	Topical gel
	Gelatin	Topical paste, emulsions, creams, oral suspensions
	Gellan gum	Ophthalmic solution
	Guar gum	Topical lotion, suspensions, oral suspensions
	Pectin	Topical paste
	Tragacanth	Oral suspensions
	Xanthan	Topical lotion, creams

Carbomer: Polyacrylic acid; Polycarbophil: Polyacrylic acid cross-linked with divinyl glycol.

4.6 CONCLUSION

Various aspects of formulating emulsions (creams and lotions), gels, ointments, and liposomes are discussed. The current thinking of the US Food and Drug Administration on the classification of semisolid dosage forms is also discussed.

REFERENCES

[1] Buhse L, Kolinski R, Westenberger B, Wokovich A, Spencer J, Chen CW, et al. Topical drug classification. Int J Pharm 2005;295:101–12.
[2] Osborne DW. Review of changes in topical drug product classification. Pharm Tech October 02, 2008;32(10).
[3] Tadros ThF. In: Tadros ThF, editor. Emulsion formation and stability. Wiley VCH, Verlag, GmbH & Co. KGaA; 2013.
[4] Kulkarni V. Liposomes in personal care products. In: Rosen MR, editor. Delivery system handbook for personal care and cosmetic products. William Andrew Publishing; 2005. pp. 285–302.
[5] Elder DP, Crowley PJ. Antimicrobial preservatives, Parts—1, 2, and 3. Am Pharm Rev January 2012.
[6] Celestino MT, Magalhães UO, Fraga AGM, Almada do Carmo F, Lione V, Castro HC, et al. Rational use of antioxidants in solid oral pharmaceutical preparations. Braz J Pharm Sci 2012;48:405–15.

CHAPTER 5

Use of Polymers and Thickeners in Semisolid and Liquid Formulations

Contents

Essential Chemistry for Formulators of Semisolid and Liquid Dosages
http://dx.doi.org/10.1016/B978-0-12-801024-2.00005-4

© 2016 Elsevier Inc.

BACKGROUND

Polymers are extremely versatile ingredients in semisolid formulations, being used to create a range of different effects, from thickening to preservation and conditioning. Polymers are macromolecules consisting of a long-chain backbone of smaller repeating units (monomers) and side groups. Polymers that contain only a single type of repeat unit are known as homopolymers, whereas polymers containing a mixture of repeat units are known as copolymers. Depending on the type of polymer, the backbone can either be linear or branched (Figure 5.1). The side groups influence the functional properties of the polymer, such as how well it thickens, conditions, and forms films.

In a solution, linear polymers can line up and interact with each other. These interactions (hydrogen bonds, van der Waals forces, and entanglements) can result in a three-dimensional network that affects viscosity and transition

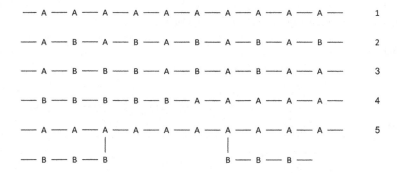

1: Homopolymer

2: Alternating copolymer – regularly alternating monomers

3: Periodic copolymer – monomers arranged in "non-regular" repeating sequence

4: Block copolymer – two or more homopolymers linked by covalent bonds

5: Graft copolymer – side chains different composition or configuration than the main chain

Figure 5.1 Polymers.

temperatures. The better the polymer chains line up, the more solid-like (crystalline) the solution behaves. Polymers that do not line up in a specific structure are called amorphous. Amorphous polymers are more commonly used in semisolid formulations than the tougher, solid-like polymers that contain microcrystalline regions.

Polymers may be either derived from naturally occurring materials (e.g., polysaccharides, proteins) or produced synthetically (by reacting monomers such as acrylic acid, vinyl pyrrolidone, or ethylene oxide). Natural materials generally need to be chemically modified before they are suitable for use in semisolid formulations.

The molecular weight of a polymer can also significantly impact its usability in semisolid formulations. Although naturally occurring polymers tend to be monodisperse (the polymer chains have uniform molecular weight), most polymers are polydisperse—the range of molecular weights being dependent on the type of polymer and reaction time. In general, the greater the molecular weight, the higher the viscosity. Highly viscous polymers can be difficult to incorporate into semisolid formulations.

5.1 POLYMERS AS STYLING AGENTS

Polymers are used extensively in hair-styling products, such as aerosols, pump sprays, gels, mousses, and styling lotions. These products have the goal of uniform application of a clear polymeric material to achieve improvements in appearance and manageability.

Hair-styling and fixative products promote the adherence between individual hair strands so that they can keep a particular shape as the polymer dries. A successful product must possess a series of properties: the hair should be held firmly in place under all conditions of humidity, but not feel too stiff; the product must dry quickly after application, but must be able to flow along the hair shaft; the hair must appear natural and glossy; the hair must be easily combed, both wet and dry; the dried product must not crack or flake when combed; it must not make the hair feel sticky; and it must be readily removed by shampooing. To achieve these properties, the polymer must not phase-separate as water permeates the system; the cohesive strength and adhesive bond of the polymer film must be less than the tensile strength and shear strength of the hair; and the film must be nonhygroscopic—otherwise it will be plasticized by absorbed water vapor under humid conditions, leading to loss of hold.

Whereas random copolymers have a solubility parameter that is intermediate between that found for homopolymers made from the same monomer units, block and graft copolymers can show dual solubility zones—each zone corresponding to the solubility parameter of the respective homopolymers. If the block copolymer is amphiphilic (i.e., comprises at least one hydrophilic polymer segment and at least one hydrophobic polymer segment), it may form films that exhibit the desired resistance to humidity but are susceptible to shampoo.

Examples of film-formers include:

PVP (polyvinyl pyrrolidone)—excellent film-former, substantive to hair, forms clear films, completely water soluble, but it absorbs water, which in humid weather can make it sticky or tacky to the touch and can cause frizz and a dull appearance to the hair. In dry weather, it can become brittle and flaky.

PVA (polyvinyl acetate)—resists absorption of water in high humidity, which leads to better hold in damp weather conditions, more flexible in dry weather so it does not flake, not as substantive to hair.

HPC (hydroxypropylcellulose)—can be used in leave-in hairstyling products such as gels, spritzes, and mousses. HPC is an excellent film-forming polymer that yields flexible, clear, thermoplastic, nontacky films.

PVP/VA copolymer (polyvinyl pyrrolidone/vinyl acetate)—a copolymer of PVP and PVA. Used to circumvent the limitations that the two polymers exhibit when used individually. Nonionic, soluble in both water and organic solvents.

Vinyl pyrrolidone/vinyl caprolactam copolymer—reduced moisture absorption and tackiness (less hydrophilic) compared to the vinyl pyrrolidone homopolymer.

Vinyl pyrrolidone/dimethylaminoethyl methacrylate copolymer—a film-forming resin and conditioning agent that provides good high-humidity curl retention, low tack, and mild substantivity to hair. Styling polymers are water and alcohol compatible and can be formulated with carbomer into clear gels. Used in gels, mousses, styling sprays (nonaerosol), styling lotions/creams, novelty stylers, and conditioners. Aids wet and dry combing. Imparts smoothness, gloss, body, and silky feeling to hair [1].

Polyurethane—an anionic hair-setting polymer for use in hair sprays with a high water content. Provides flexible and elastic hold with excellent curl retention and high humidity resistance.

Acrylic copolymers—low-molecular-weight, high-viscosity copolymers, soluble in alkaline water.

5.2 POLYMERS AS CONDITIONING AGENTS

Polymers are used extensively in cosmetic formulations to improve the surface of hair or skin. These conditioning polymers are used in shampoos and conditioners to provide slip, detangling, and antistatic properties, and to add body, gloss, shine, firmness, and texture. In skin creams, polymers provide moisturization and humectancy. Typically, these conditioning agents are cationic polymers, proteins, or silicones.

5.2.1 Cationic Polymers

- Cationic Polymers are linear or branched polymers that have been chemically modified to have positively charged sites throughout the polymeric structure.
- Quaternary ammonium compounds (quats) have four chemical groups attached to the nitrogen of the ammonium group (hydrogen groups replaced by alkyl groups), and a chlorine or bromine ion at the end of the molecule. Quats have been used extensively as conditioning agents.
- Although some quats are obtained via chemical modifications of naturally derived polymers such as guar gum and cellulose, most are completely synthetic.
- Quats have a long hydrophobic C–H tail and a positively charged (cationic) N-group that is attracted to the negatively charged (anionic) proteins of the hair or skin. The electrostatic interaction, coupled with the fatty nature of the long C–H tail, inhibits rinse-off and makes the hair cuticle smooth, soft, and lubricious. The longer the fatty chain, the more lubricant the quat.
- Many cationic polymers are compatible with anionic surfactants. In a shampoo formula, cationic polymers rely on a dilution/deposition mechanism. When applied to the hair, the shampoo becomes diluted; this causes the concentration of the anionic surfactant to fall below the critical micelle concentration, which causes the polymer to precipitate out on the hair [2]. When the shampoo is rinsed away, some of the polymer remains on the hair because of its insolubility in water and electrostatic interaction.
  Examples of cationic polymers:
- Polyquaternium-10 (quaternized hydroxyethyl cellulose. Antistatic agent; film former; hair fixative)
- Polyquaternium-7 (copolymer of acrylamide and diallyldimethylammonium chloride. Antistatic agent; film former; hair fixative)

- Polyquaternium-11 (copolymer of vinylpyrrolidone and quaternized dimethylaminoethyl methacrylate. Antistatic agent; film former; hair fixative)
- Guar hydroxypropyltrimonium chloride (water-soluble quaternary ammonium derivative of guar gum. It gives conditioning properties to shampoos and after-shampoo hair care products)
- Behentrimonium chloride (an antistatic agent and a hair conditioner and also a preservative)

5.2.2 Silicones

- Produced by polymerization of siloxanes. They are mixed inorganic–organic polymers with the chemical formula $[R_2SiO]_n$. These materials consist of an inorganic silicon–oxygen backbone ($\cdots$–Si–O–Si–O–Si–O–$\cdots$) with organic side groups attached to the silicon atoms.
- Silicones provide a protective layer on the hair shaft and also provide shine. They reduce combing friction, have an emollient effect (on both skin and hair), impart gloss, and reduce static charge between hair strands. Examples of silicones:
- Dimethicone (polydimethylsiloxane) (Figure 5.2)—widely used in shampoos, conditioners, and skin creams. Linear polymer with pendant methyl groups. Insoluble in water. In creams, provides smoothness, lubricity, and shine without being oily. In hair care, improves combing, shine, and manageability
- Dimethicone copolyols—dimethicone copolymerized with polyethylene glycol and polypropylene glycol. Improves water solubility of dimethicone, and offers surfactant qualities
- Amodimethicone—amine functionalized; provides improved compatibility, feel, softness, and reduces frizz
- Decamethylcyclopentasiloxane (D5) (Figure 5.3)—used in deodorants, sunblocks, sunless tanning lotion, hair sprays and conditioners, and skin

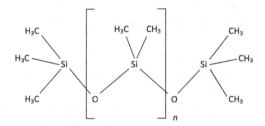

Figure 5.2 Dimethicone.

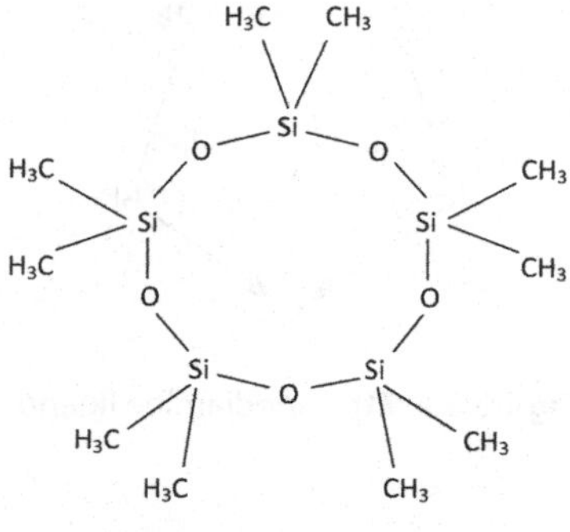

Figure 5.3 Decamethylcyclopentasiloxane.

care products. It makes the hair easier to brush without breakage. It is also used as part of silicone-based personal lubricants and is an emollient

5.2.3 Proteins

- Derived from both plant (wheat, rice, soy, corn, vegetables) and animal (milk, collagen, keratin) sources, they are hydrolyzed to improve water solubility.
- They are naturally absorbed due to the similarity of their proteinaceous structure to hair and skin. They are nonocclusive film formers that bind water and enhance the ability of the skin to absorb and retain moisture.
- They strengthen the hair via molecular cross-linking.

5.3 POLYMERS AS PRESERVATIVES [3,4]

A preservative is a naturally occurring or synthetically produced substance that is added to a product to prevent decomposition by microbial growth or by undesirable chemical changes (including antioxidants and chelators). Ideally, preservatives should be heat stable, work over a wide range of pH, and be soluble in water.

Antimicrobial preservatives have biocidal properties that inhibit the growth of bacteria, yeast, and molds. They are added to semisolid formulations to protect them from contamination by microorganisms present in the air, water, and on our own skin, thereby increasing the product's shelf life and consumer safety. A list of preservatives allowed for use in cosmetic products is published by the European Commission (Annex V of the cosmetic ingredients database "ec.europa.eu/consumers/cosmetics/cosing") and International Nomenclature of Cosmetic Ingredients directory. Commonly used antimicrobial preservatives comprise both "small molecules" (e.g.,

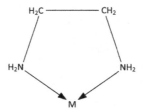

Figure 5.4 Ethylenediamine ligand.

benzyl alcohol, parabens, phenoxyethanol) and polymers (e.g., benzalkonium chloride, polyaminopropyl biguanide, and alkyl (C_{12}–C_{22}) trimethyl ammonium bromide and chloride). Frequently, blends and combinations of preservatives are used to give broader spectrum protection or to improve stability. Standardized tests to assess the antimicrobial effectiveness of preserved formulations have been published [5].

Antioxidants are added to protect oxygen-sensitive formulation ingredients from oxidation. They are also used in cosmetic formulations to counter free radicals that damage lipids and proteins. Examples of commonly used "small molecule" antioxidants are butylated hydroxytoluene, butylated hydroxyanisole, retinol, ascorbic acid, and isoflavones; polymeric antioxidants include açai oil, coenzyme Q-10, and polyphenols.

Chelators are small molecules that bind very tightly to metal ions, making the metal ion chemically inert. Chelation involves the formation of a coordination compound in which a metal atom or an ion is bound to a ligand at two or more points on the ligand, so as to form a heterocyclic ring containing the metal atom (Figure 5.4). The bonding between a metal and a ligand generally involves formal donation of one or more of the ligand's electron pairs. The nature of metal–ligand bonding can range from covalent to ionic. Although most chelators used in semisolid formulations are "small molecules" (e.g., ethylenediaminetetraacetic acid, citric acid, hydroxyquinolone, salts of tartaric acid), some complex proteins and polymers also have chelating properties [6] (e.g., transferrin, amino acids, cyclams, hydroxamic acid derivatives).

5.4 POLYMERS AS PENETRATION ENHANCERS

For semisolid formulations to "deliver active ingredients", they must first penetrate the skin. This process can be rate limiting, and has been described as a series of consecutive steps [7,8]: diffusion of chemical from the formulation to the surface of the skin; partitioning into and diffusing through the stratum corneum, epidermis, and dermis; before finally partitioning into fat deposits or being redistributed via blood capillaries. Chemical penetration enhancers have been used to reversibly reduce the barrier function of the stratum corneum, leading to an increased systemic availability of drugs [9].

Examples of polymeric chemical penetration enhancers:

- Oleic acid—a monounsaturated fatty acid that increases the partitioning of drugs into the stratum corneum by disrupting the barrier function of the skin. It increases the solubility of lipophilic drugs by forming lipophilic complexes. It enhances the delivery of both lipophilic and hydrophilic drugs.
- Azone (laurocapram) [10]—disrupts the lipid bilayers and increases the fluidity and permeation in the lipid regions of the skin. Also fluidizes the hydrophobic regions of the lamellate structure. It enhances penetration of both lipophilic and hydrophilic drugs.
- Dendrimers—repetitively branched molecules, typically symmetric around the core, often adopt a spherical three-dimensional morphology. They have been used to encapsulate hydrophobic drugs. They interact with lipids and keratin in the stratum corneum and increase drug partitioning.
- Monoolein—a monoglyceride with a structure similar to oleic acid. A polar lipid that is insoluble in water. Forms a bicontinuous cubic liquid crystalline phase that aids drug delivery. It disrupts the lamellar structure of the bilayers in the stratum corneum, thereby increasing lipid fluidity. It also solubilizes lipophilic compounds in the skin.
- Oxazolidinones—high-molecular-weight compounds with structural features closely related to the sphingosine and ceramide lipids found in the upper skin layers. They fluidize the bilayers in the stratum corneum.
- SEPA (2-n-nonyl-1,3-dioxolane)—fluidizes lipids and alters the structure of the stratum corneum
- Phospholipids—formulated into microemulsions, vesicles, and micellar systems. They disrupt the ordered bilayer structure in the stratum corneum and enhance the partitioning of encapsulated drugs.
- Surfactants—surface-active agents that alter the barrier function of the stratum corneum by removing water-soluble components and emulsifying sebum. This degreasing effect can cause irritation, especially with cationics. Examples are:
  - Sodium lauryl sulfate (anionic)
  - Cetyltrimethyl ammonium bromide (cationic)
  - Polyoxyethylene sorbitan monopalmitate (nonionic)
  - Lauryl betaine (amphoteric/zwitterionic).

5.5 POLYMERS AS THICKENING AND SUSPENDING AGENTS [11]

5.5.1 Emulsions

An emulsion is a thermodynamically unstable two-phase system consisting of at least two immiscible liquids (e.g., oil and water), one of which is dispersed in the form of small droplets throughout the other. Surfactants (emulsifiers) are used to stabilize emulsions by reducing the interfacial tension between the oil and water phases (surface tension theory), and/or by creating a film over the dispersed-phase droplets, which repel each other (repulsion theory). Polymeric thickeners are commonly used to stabilize emulsions by increasing the viscosity of the continuous phase, thereby helping to maintain the suspension of dispersed droplets. Polymeric thickeners tend to be less affected by pH and salt content than surfactant thickeners and require the use of less material to get the same thickening properties.

Suspending agents (e.g., Carbopol® polymers) prevent solid material in emulsions (e.g., pearling materials, pigments, antiperspirants, inorganic sunblocks) from depositing as sediment.

5.5.2 Gels

A gel consists of suspended particles that create a three-dimensional structure of interlacing particles or solvated macromolecules that restrict the movement in the dispersing medium. These gelling agents undergo a high degree of cross-linking or association when hydrated, thereby increasing the viscosity. Gels tend to be shear thinning (spread easily with applied pressure or friction) water, or water/alcohol, solutions thickened with a gelling agent or combination of gelling agents. The chemistry of the gelling agents varies with the specific application.

In addition to water-based gels, anhydrous gels thickened with organo-modified clays have been cited in the US and international patents [12].

5.5.2.1 Examples of Gelling Agents

- Natural hydrocolloids: tragacanth, guar gum, carrageenan, xanthan gum, locust bean gum, gum acacia, alginates, gelatin
- Cellulose and derivatives: hydroxyethylcellulose, hydroxymethylcellulose, methylcellulose, carboxymethylcellulose, ethylcellulose, hydroxypropylcellulose
- Synthetic polymers: carbomers (Carbopol® polymers), poloxamers (Pluronics®), polyvinyl alcohol
- Clays: bentonite, magnesium aluminum silicate (Veegum®)

5.5.3 Suspensions

A suspension is a two-phase system consisting of finely divided solid material dispersed in a liquid. The primary disadvantage of a suspension is one of physical stability, that is, that the suspended material tends to settle over time. Suspending agents are insoluble particles that are dispersed in the liquid vehicle to reduce this rate of sedimentation. Most suspending agents perform two functions: they form films around the suspended particles, decreasing interparticle attraction; they also act as thickening agents, increasing the viscosity of the solution. Suspending agents help reduce the sedimentation rate of particles in a suspension in accordance with Stokes' Law (Eqn (5.1)):

$$v = 2 \left(d_1 - d_2 \right) gr^2 / 9\eta \tag{5.1}$$

in which v is the settling velocity, r is the radius of the suspended particle, η is the viscosity of the liquid vehicle, d_1 is the density of the suspended particle, d_2 is the density of the liquid vehicle, and g is the gravitational constant.

A good suspension should have well-developed thixotropy (shear thinning). At rest (low-shear rates), the solution should be sufficiently viscous to prevent sedimentation of the particles. When agitation is applied (during shaking, pouring, or spreading), the viscosity should decrease to provide good flow characteristics. Oral suspensions and syrups are examples of "thickened" liquid dosage forms that may be found desirable by patients who have difficulty in swallowing tablets or capsules—for example, very young children.

5.5.3.1 Examples of Suspending Agents

Natural hydrocolloids: gum acacia, tragacanth, alginates, carrageenan, locust bean gum, guar gum, gelatin, xanthan gum, powdered cellulose, microcrystalline cellulose (MCC).

Semisynthetic hydrocolloids: methylcellulose, sodium carboxymethylcellulose, hydroxyethylcellulose, hydroxypropylmethylcellulose, carboxymethylcellulose.

Synthetic hydrocolloids: carbomers (Carbopol® polymers)

Clays: bentonite, magnesium aluminum silicate (Veegum®).

5.6 CHARACTERISTICS OF COMMONLY USED THICKENING, GELLING, AND SUSPENDING AGENTS

5.6.1 Natural Hydrocolloids

5.6.1.1 Agar

- Agar is a polymer, obtained from algae, made up of subunits of the sugar galactose. It is a mixture of two components: the linear polysaccharide

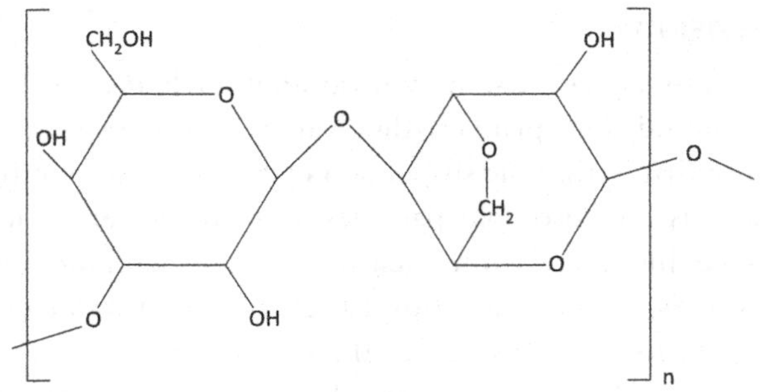

Figure 5.5 "Agarose".

agarose and a heterogeneous mixture of smaller molecules called agaro-pectin [13] (Figure 5.5).

5.6.1.2 Alginates

- Alginate is an anionic polysaccharide extracted from brown seaweed [14], composed of mannuronic acid and glucuronic acid monomers.
- In practice, alginate is used at concentrations less than 10% w/w to avoid pourability problems by too much of an increase in the viscosity of suspension.
- Cold-water soluble and cold setting. Heat and freeze/thaw stable. Broad range of flow properties for aqueous-based systems.
- Maximum viscosity is observed at a pH range of 5-9. Fresh solution has the highest viscosity, after which viscosity gradually decreases and acquires a constant value after 24 h.
- Alginate solutions lose viscosity when heated above 60 °C due to depolymerization.
- Alginate salts have the same suspending action as that of Tragacanth (Figure 5.6).

5.6.1.3 Carrageenan

- Carrageenans are polysaccharides that are extracted from red edible sea-weeds. All carrageenans are high-molecular-weight polysaccharides made up of repeating galactose units and 3,6-anhydrogalactose (3,6-AG), both sulfated and nonsulfated. The units are joined by alternating α-1–3 and β-1–4 glycosidic linkages.
- Three main commercial classes of carrageenan: kappa (Figure 5.7), iota, and lambda. All are soluble in hot water, but in cold water only the lambda form (and the sodium salts of the other two) is soluble.

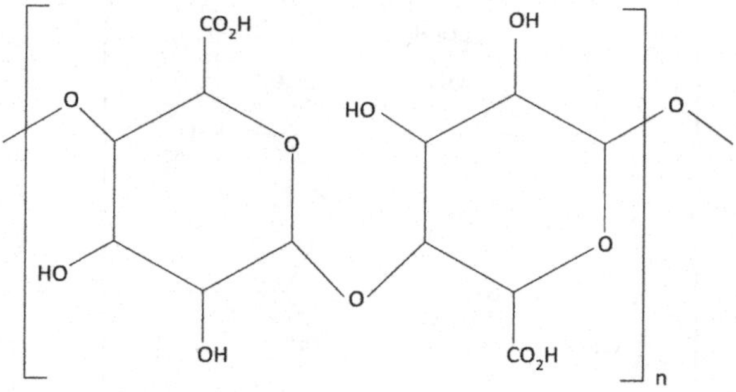

Figure 5.6 Alginic acid.

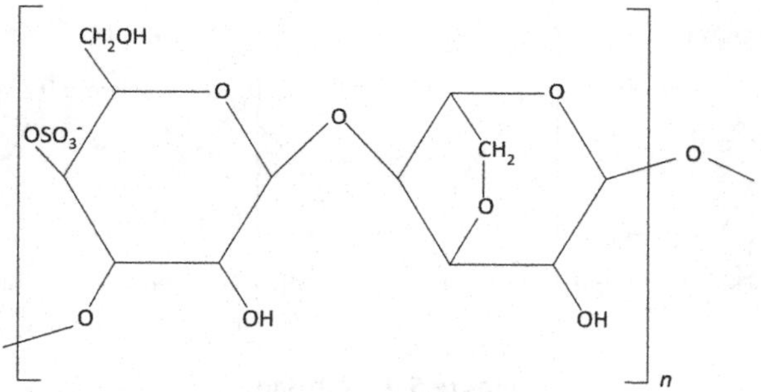

Figure 5.7 Kappa carrageenan.

- Suspending agent for insoluble particles through extended storage. Gelation capabilities at ambient and refrigerator temperatures (kappa forms strong, rigid gels in the presence of potassium ions; iota forms soft gels in the presence of calcium ions). Controlled flow properties [14].
- Thickening and/or gelling agent. Lambda carrageenan acts as a thickening agent that does not gel.
- Carrageenan is stable at a pH range of 6-10.

5.6.1.4 Chitin and chitosan
- Chitin is a long-chain polymer of an *N*-acetylglucosamine, a derivative of glucose, and is found in many places throughout the natural world. It is the main component of the cell walls of fungi, the exoskeletons of arthropods such as crustaceans (e.g., crabs, lobsters, and shrimps) and insects, the radulas of mollusks, and the beaks and internal shells of cephalopods, including squid and octopuses.

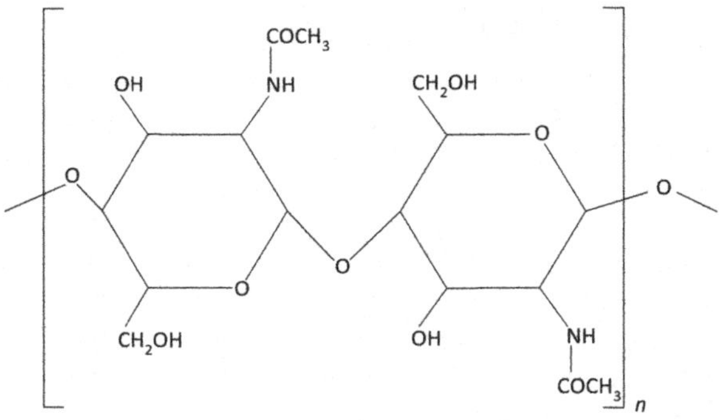

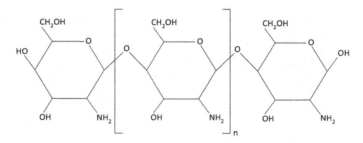

Figure 5.8 Chitin.

Figure 5.9 Chitosan.

- Chitin may be described as cellulose with one hydroxyl group on each monomer replaced with an acetyl amine group. This allows for increased hydrogen bonding between adjacent polymers, giving the chitin–polymer matrix increased strength.
- Chitosan is produced commercially by deacetylation of chitin (Figures 5.8 and 5.9).

5.6.1.5 Gelatin
- Gelatin is a mixture of peptides and proteins produced by partial hydrolysis of collagen extracted from the skin, bones, and connective tissues of animals.
- Gelatin forms a solution of high viscosity in water, which sets forming a semisolid colloidal gel on cooling. Gelatin is also soluble in most polar solvents.

5.6.1.6 Guar gum
- Guar gum is primarily the ground endosperm of guar beans. It is a polysaccharide composed of the sugars, galactose and mannose. The

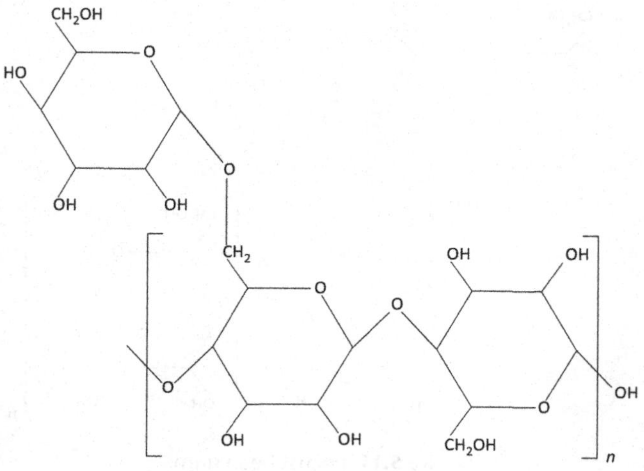

Figure 5.10 Guar gum.

backbone is a linear chain of β-1,4-linked mannose residues to which galactose residues are 1,6-linked at every second mannose, forming short side branches.

- Aqueous solubility and thickening capability decrease sharply below pH 4.5.
- Guar gum is more soluble than locust bean gum and is a better stabilizer, as it has more galactose branch points. Unlike locust bean gum, it is not self-gelling [15]. In water, it is nonionic and hydrocolloidal. It is not affected by ionic strength, but will degrade at pH extremes at temperature (e.g., pH 3 at 50 °C) [14]. It remains stable in solution over the pH range 5–7. Strong acids cause hydrolysis and loss of viscosity, and alkalies in strong concentration also tend to reduce viscosity.
- Guar gum is used as a suspending agent (Figure 5.10).

5.6.1.7 Gum acacia/gum arabic

- Acacia is a natural gum made of hardened sap taken from two species of the acacia tree: *Senegalia senegal* and *Vachellia seyal*. It is a complex mixture of glycoproteins and polysaccharides.
- A binder, emulsifying agent, and a suspending or viscosity-increasing agent [16]. Not a good thickening agent.
- For dense powders, acacia alone is not capable of providing suspending action, therefore it is mixed with Tragacanth, starch, and sucrose, which is commonly known as Compound Tragacanth Powder BP.

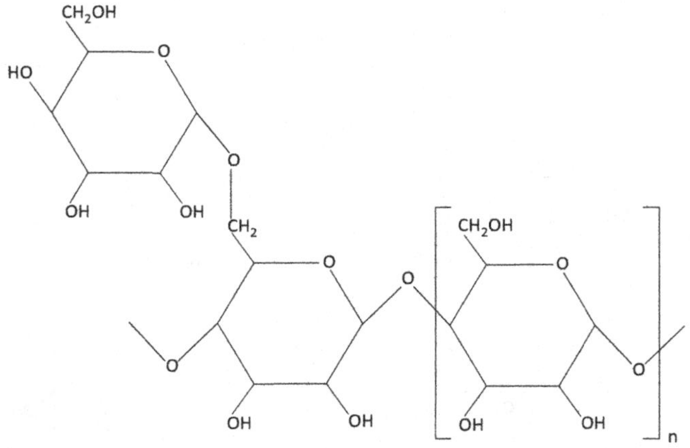

Figure 5.11 Locust bean gum.

5.6.1.8 Locust bean gum

- Locust bean gum is a galactomannan vegetable gum extracted from the seeds of the carob tree. It consists chiefly of high-molecular-weight hydrocolloidal polysaccharides, composed of galactose and mannose units combined through glycosidic linkages.
- Locust bean gum is dispersible in either hot or cold water, forming a sol having a pH between 5.4 and 7.0, which may be converted to a gel by the addition of small amounts of sodium borate.
- Thickening agent, gelling agent.
- In locust bean gum, the ratio of mannose to galactose is higher than in guar gum, giving it slightly different properties, and allowing the two gums to interact synergistically so that together they make a thicker gel than either one alone.
- Locust bean gum significantly improves gel strength and texture and prevents syneresis when used in combination with carrageenans.
- Its synergy with xanthan gum provides highly elastic gels with very limited syneresis [17] (Figure 5.11).

5.6.1.9 Pectin

- Pectin is a structural heteropolysaccharide contained in the primary cell walls of terrestrial plants. Pectins are rich in galacturonic acid.
- Pectin is used as a gelling agent, thickening agent (Figure 5.12).

5.6.1.10 Starch

- Starch is a carbohydrate consisting of a large number of glucose units joined by glycosidic bonds. It consists of two types of molecules: the

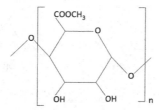

Figure 5.12 Pectin.

linear and helical amylose and the branched amylopectin. This polysaccharide is produced by most green plants.

- Starch is insoluble in cold water or alcohol.
- Starch becomes soluble in water when heated. The granules swell and burst, the semicrystalline structure is lost, and the smaller amylose molecules start leaching out of the granule, forming a network that holds water and increases the mixture's viscosity (Figure 5.13).

5.6.1.11 Tragacanth

- Tragacanth is a natural gum obtained from the dried sap of several species of Middle Eastern legumes of the genus *Astragalus* [18]. It is a viscous, odorless, tasteless, water-soluble mixture of polysaccharides.
- Tragacanth provides thixotrophy to a solution (forms pseudoplastic solutions). The maximum viscosity of the solution is achieved after several days, due to the time taken to hydrate completely.
- Tragacanth is stable at a pH range of 4-8.
- It is a better thickening agent than acacia.
- Tragacanth is used as a suspending agent, emulsifier, thickener, and stabilizer.

5.6.1.12 Xanthan Gum

- Xanthan gum is a polysaccharide secreted by the bacterium *Xanthomonas campestris* [18]. It is composed of pentasaccharide repeat units, comprising glucose, mannose, and glucuronic acid in the molar ratio 2:2:1 [19].
- Rheology modifier. A stabilizer (in cosmetic products, e.g., to prevent ingredients from separating).
- Hot and cold water soluble. Gives neutral pH, pseudoplastic (shear thinning) solutions.
- Very stable under a wide range of temperatures and pH (3–12).
- Xanthan gum is used as a suspending agent (Figure 5.14).

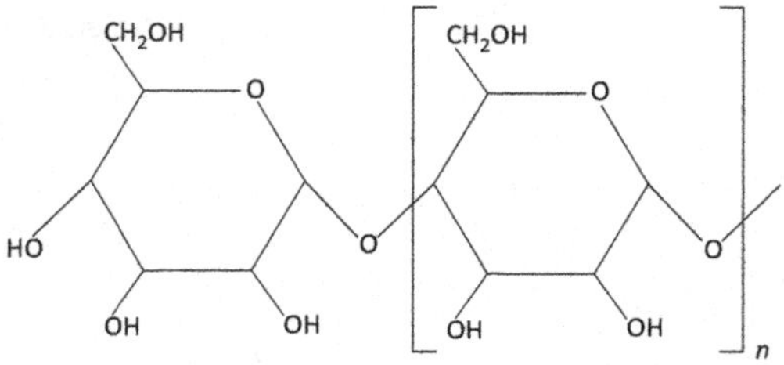

Amylose

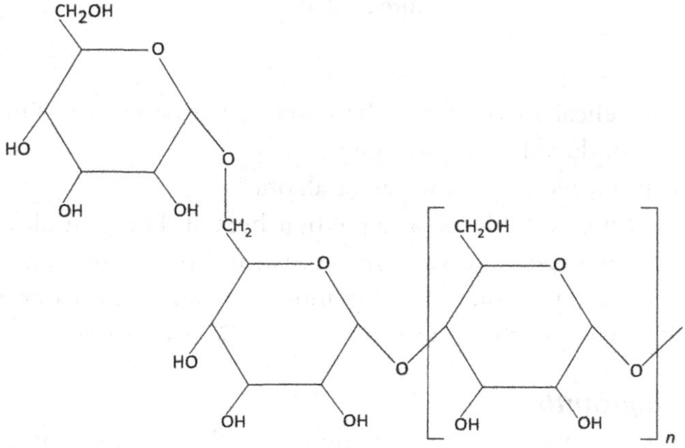

Amylopectin

Figure 5.13 Starch.

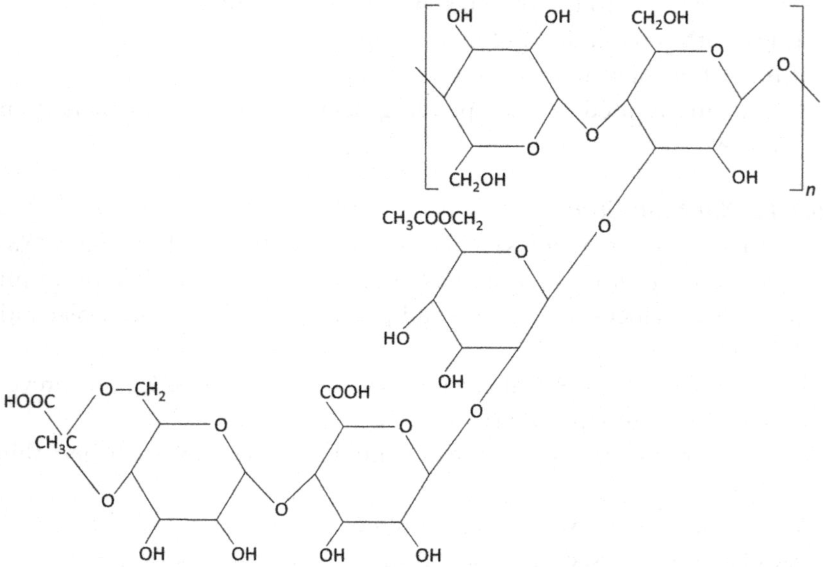

Figure 5.14 Xanthan gum.

5.6.2 Cellulose and Derivatives

5.6.2.1 Cellulose/Microcrystalline Cellulose (MCC)

- Cellulose is an important structural component of the primary cell wall of green plants and many forms of algae. It is a polysaccharide consisting of a linear chain of several 100 to over 10,000 $\beta(1\rightarrow4)$ linked D-glucose units [20]. These linear cellulose chains are bundled together as microfibrils, spiraled together in the walls of plant cells. Each microfibril exhibits a high degree of three-dimensional internal bonding resulting in a crystalline structure that is insoluble in water and resistant to reagents. Relatively weak segments of the microfibril with weaker internal bonding exist, called amorphous regions. The crystalline region is isolated to produce MCC (Figure 5.15).
- Cellulose is an effective thickener in water-based systems. It is usually used in high water content systems (>90%). It is compatible with anionic and cationic surfactants, electrolytes, and nonionic materials. It builds viscosity because of its high molecular mass. It exhibits substantial shear thinning [21].
- MCC is not soluble in water, but it readily disperses to give thixotropic gels. It is used in combination with sodium carboxy methyl cellulose (NaCMC), methyl cellulose (MC), or hydroxypropyl methyl cellulose (HPMC), because they facilitate dispersion of MCC.
- MCC–MC is stable at a pH range of 1-11.

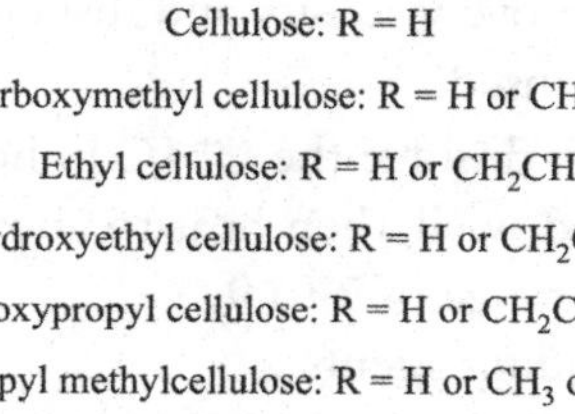

Cellulose: R = H

Carboxymethyl cellulose: R = H or CH_2CO_2H

Ethyl cellulose: R = H or CH_2CH_3

Hydroxyethyl cellulose: R = H or CH_2CH_2OH

Hydroxypropyl cellulose: R = H or $CH_2CH(OH)CH_3$

Hydroxypropyl methylcellulose: R = H or CH_3 or $CH_2CH(OH)CH_3$

Methyl cellulose: R = H or CH_3

Figure 5.15 Cellulosics.

- MCC–alginate complex compositions are excellent suspending agents. Suspensions prepared with them are redispersible with a small amount of agitation and maintain viscosity even under high-shear environments.
- Mixtures of MCC–NaCMC are available from FMC Corp. under trademark Avicel®. In water, with shear, it forms a three-dimensional matrix comprised of millions of insoluble microcrystals that form an extremely stable, thixotropic gel. Avicel® MCC functions at any temperature and provides superior freeze/thaw and heat stability to finished products.
- Avicel® RC-591 gels are stable at pH 4–11. They provide a slow uniform settling rate, suspension stability, and are readily flocculated by small amounts of electrolytes/cationic polymers/surfactants. The use of protective colloids (xanthan, MC, CMC at 15 to 20%) can prevent flocculation.
- Avicel® RC-591: suspending agent, emulsion stabilizer [14]. Used in suspension products in which shelf-life stability (e.g., no physical separation) is desired.
- Avicel® CL-611 is recommended for reconstitutable suspensions.
  Carboxymethyl Cellulose (CMC)
- CMC (or cellulose gum) is a cellulose derivative with carboxymethyl groups ($-CH_2-COOH$) bound to some of the hydroxyl groups of the glucopyranose monomers that make up the cellulose backbone. It is synthesized by the alkali-catalyzed reaction of cellulose with chloroacetic acid. The polar (organic acid) carboxyl groups render the cellulose soluble and chemically reactive. It is often used as its sodium salt, sodium carboxymethyl cellulose.
- CMC is available at different viscosity grades: low, medium, and high. The choice of CMC grade is dependent on the viscosity and stability of the suspension.
- In the case of high-viscosity CMC, the viscosity significantly decreases when the temperature rises to 40 °C from 25 °C, and may cause product stability concerns. To improve viscosity and suspension stability, medium viscosity CMC is widely used.
- When formulating, either dissolve the CMC in hot water, or disperse in cold water to hydrate and swell, then heat to about 60 °C to gel. Maximum stability is at a pH range of 7 to 9.
- CMC is used as a suspending agent.
  Sodium Carboxymethyl Cellulose (NaCMC)
- NaCMC is available in various viscosity grades depending on the extent of polymerization. It is made by reacting sodium monochloroacetate with alkali-cellulose [1].

- It is an anionic water-soluble polymer, soluble in both hot and cold water. It is stable over a pH range of 5-10 and forms slightly thixotropic solutions. It is incompatible with polyvalent cations.
- In addition to thickening aqueous systems, NaCMC is used in personal care products for water binding, syneresis control, and its ability to suspend pigments and active ingredients in solution.
- NaCMC is available from Ashland (Aqualon™ [1]) and The Dow Chemical Company (WALOCEL™ binder [22]).

Ethyl Cellulose (EC)

- EC is a derivative of cellulose in which some of the hydroxyl groups on the repeating glucose units are converted into ethyl ether groups.
- It is insoluble in glycerin, propylene glycol, or water. Soluble in many organic solvents including ethanol and natural oils.
- EC forms thermoplastic polymers, and is available from The Dow Chemical Company (ETHOCEL™ polymer [22]).
- EC is used as a film former, fragrance stabilizer, and thickener for perfumes and body creams (waterproof sunscreens).

Hydroxyethyl Cellulose (HEC)

- HEC is a nonionic, water-soluble polymer. It is made by reacting ethylene oxide with alkali-cellulose [1]. Solutions of HEC are pseudoplastic or shear thinning.
- HEC is soluble in cold or hot water giving crystal-clear solutions of varying viscosities. Low- to medium-molecular-weight types are fully soluble in glycerol and have good solubility in hydroalcoholic systems containing up to 60% ethanol. HEC is generally insoluble in organic solvents.
- HEC is stable at a pH range of 2-12.
- HEC is used as a suspending agent.

Hydroxypropyl Cellulose (HPC)

- HPC is an ether of cellulose in which some of the hydroxyl groups in the repeating glucose units have been hydroxypropylated forming-$OCH_2CH(OH)CH_3$ groups using propylene oxide.
- HPC has a combination of hydrophobic and hydrophilic groups. It is readily soluble in cold water, alcohols, many polar organic solvents, polyethylene glycol, and propylene glycol. HPC is generally insoluble in water over 105 °F (40 °C); however, this precipitation phenomenon occurs only in water and is fully reversible upon cooling. High-molecular-weight grades (high-viscosity types) are effective thickeners and film formers, whereas lower-molecular-weight grades are often employed for their excellent film-forming properties [1].

™ WALOCEL and ETHOCEL are trademarks of The Dow Chemical Company.

- HPC is stable at a pH range of 6-8.
- HPC is used as a thickener for aqueous and polar organic systems, a suspending agent, and as an emulsion stabilizer.

Hydroxypropyl methylcellulose (HPMC)

- Propylene oxide is used in addition to methyl chloride to obtain hydroxypropyl substitution ($-OCH_2CH(OH)-CH_3$) on the anhydro-glucose units of the cellulose backbone [1,22].
- HPMC is an inert, viscoelastic, nonionic polymer, quite hydrophilic and readily soluble in cold water. HPMC exhibits a thermal gelation property. As the water temperature is raised, the polymer passes through a thermal gelation phase, in which solution viscosity rises rapidly forming a nonflowable but semiflexible mass with thixotropic behavior, followed by flocculation and precipitation of the polymer out of solution. This precipitation is thermally reversible and occurs at different temperatures, depending on the type and degree of substitution.
- HPMC is compatible with water and alcohol, disperses in cool water, and is stable at a pH range of 3-11.
- HPMC is used as a thickening and suspending agent.

Methyl Cellulose (MC)

- MC is produced by heating cellulose with caustic solution (e.g., a solution of sodium hydroxide) and treating it with methyl chloride, to replace the hydroxyl residues ($-OH$ functional groups) with methoxide ($-OCH_3$) groups.
- It is available in several viscosity grades resulting from different methylation and polymer chain lengths [1,22].
- MC dissolves in cold (but not in hot) water, forming a clear viscous solution or gel. These are best prepared by dispersing the powder in hot water, then cooling this dispersion down while stirring.
- As MC is nonionic, it is compatible with many ionic adjuvants. It is stable at a pH range of 3-11.
- MC is used as a thickener and suspending agent.

5.6.3 Synthetic Polymers

Carbomers (Carbopol® polymers)

- Carbomers are high-molecular-weight homo- and copolymers of acrylic acid cross-linked with a polyalkenyl polyether [23,1]. They are anionic in nature and acidic in their unneutralized state and have to be neutralized with an appropriate base to achieve their thickening ability (Figure 5.16).

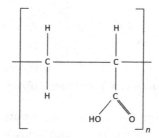

Figure 5.16 Carbomer.

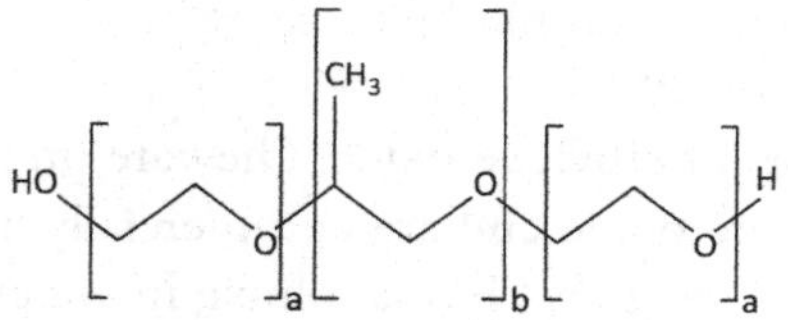

'a' represents ethylene oxide portion,
'b' represents propylene oxide portion

Figure 5.17 Poloxamer.

- Neutralizing with inorganic bases gives stable water soluble gel. Use of triethanolamine forms gels that can tolerate high-alcohol concentrations.
- Application across broad pH range (4.5–10). Used in aqueous, anhydrous, hydroalcoholic systems.
- Not heat sensistive—formulations can be autoclaved.
- Possess shear thinning (pseudoplastic) properties. Sensitive to electrolytes, multivalent cations. Use propylene glycol or glycerine to increase viscosity. Use electrolytes to decrease viscosity.
- Carbomers are used as thickeners, emulsion stabilizers, gel formers, and suspending agents.
  Poloxamers (Pluronics®) [24]
- Poloxamers are copolymers of polyoxyethylene and polyoxypropylene. They are nonionic triblock copolymers composed of a central hydrophobic chain of polyoxypropylene (poly(propylene oxide)) flanked by two hydrophilic chains of polyoxyethylene (poly(ethylene oxide)) (Figure 5.17).
- Aqueous solutions of poloxamer are very stable in the presence of acids, alkalis, and metal ions. Poloxamers are readily soluble in aqueous, polar, and nonpolar organic solvents [25].

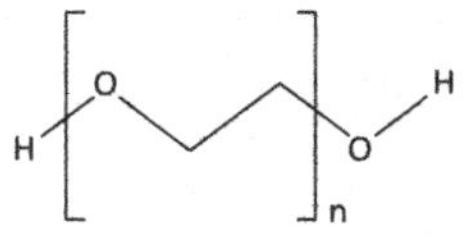

Figure 5.18 Polyethylene glycol.

- They form thermoreversible gels at concentrations of 15-50% that are liquid at cool temperatures, gels at room/body temperatures [26–28].
- Because of their amphiphilic structure, the polymers have surfactant properties and can be used to increase the water solubility of hydrophobic, oily substances or otherwise increase the miscibility of two substances with different hydrophobicities.

Polyethylene Glycol (PEG)

- PEGs are polymers of ethylene oxide. They are prepared by the polymerization of ethylene oxide and are commercially available over a wide range of molecular weights. PEG is soluble in water, methanol, ethanol, acetonitrile, benzene, and dichloromethane, and is insoluble in diethyl ether and hexane (Figure 5.18).
- They are used in a wide range of cosmetic and personal care preparations including creams, lotions, sticks, cakes, powders, gels, and aerosols. When used in lotion-type preparations, they solubilize actives, as well as add a "silky feel" without greasiness. They are water-soluble, odorless, neutral, lubricating, nonvolatile, and nonirritating and make excellent coupling agents, solvents, vehicles, humectants, lubricants, binders, and bases.
- PEGs make excellent water-soluble ointment bases; they spread easily and evenly over the skin even if the skin is moist. Their good water solubility makes it easy to incorporate aqueous ingredients into the formulation, and they do not become rancid or have a nutritional value to support microbial growth.
- PEGs are used as humectants and viscosity modifiers [22].

Polypropylene Glycol (PPG)

- PPG is a polymer of propylene glycol. It has many properties in common with polyethylene glycol. The polymer is a liquid at room temperature. Solubility in water decreases rapidly with increasing molar mass (Figure 5.19).

Polyvinyl Alcohol (PVA)

- PVA is a water-soluble synthetic polymer made by dissolving polyvinyl acetate (PVAc) in an alcohol such as methanol and treating it with an alkaline catalyst such as sodium hydroxide. When the reaction is allowed to proceed to completion, the product is highly soluble in water and

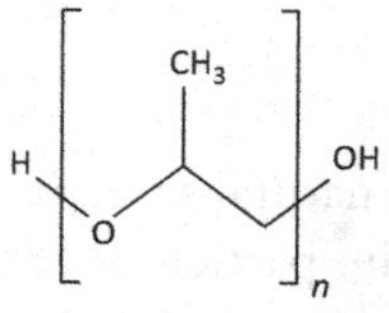

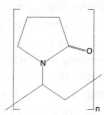

Figure 5.19 Polypropylene glycol.

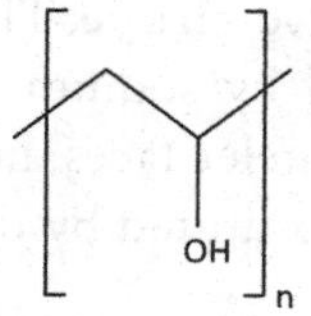

Figure 5.20 Polyvinyl alcohol.

Figure 5.21 Polyvinyl pyrrolidone.

insoluble in practically all organic solvents. Incomplete removal of the acetate groups yields resins less soluble in water and more soluble in certain organic liquids (Figure 5.20).

- PVA exhibits crystallinity and has excellent film forming, emulsifying, and adhesive properties. It is also used as a water-soluble protective film. Polyvinyl Pyrrolidone (PVP)
- PVP is an amorphous polymer that forms complexes with hydrogen donors, such as phenols and carboxylic acids, as well as anionic dyes and inorganic salts [1]. It is a water-soluble nonionic polymer made from the monomer N-vinylpyrrolidone [29]. It is also soluble in other polar solvents (Figure 5.21).
- PVP is temperature resistant and pH stable.
- PVP is used as a film former, thickening agent, and suspension stabilizer [30]. It is also used as an aid for increasing the solubility of drugs in liquid and semiliquid dosage forms (syrups, soft gelatine capsules) and as an inhibitor of recrystallization.

5.6.4 Clays

Bentonite

- Bentonite is an absorbent natural smectite clay. It has a colloidal structure in water. Each smectite particle is composed of thousands of submicroscopic platelets stacked in a sandwich fashion with a layer of water between each. A single platelet is 1 nm thick and up to several 100 nm across. The faces of these platelets carry a negative charge, whereas edges have a slightly positive charge. The net negative charge of the platelet is mostly balanced by sodium ions. These charge-balancing ions are associated with platelet faces and are termed "exchangeable" because they are readily substituted by other cations.
- It is stable at $pH > 6$.
- Most often, bentonite suspensions are thixotropic (shear thinning), although rare cases of rheopectic (shear thickening) behavior have also been reported [31]. At higher concentrations, bentonite suspensions begin to take on the characteristics of a gel (a fluid with minimum yield strength required to make it move).
- Bentonite is used as a suspending and rheological agent.
  Magnesium aluminum silicate (Veegum®)
- Veegum® is smectite clay that forms a colloidal structure in water that is well suited for suspension stabilization. It exhibits synergistic interaction with thickening agents [32]:
  - Anionic naturally derived thickeners generally show the best compatibility and synergy.
  - Xanthan gum enhances the flow properties, suspension efficiency, and the electrolyte, acid, and alkaline compatibility of smectite clay dispersions.
  - NaCMC is strongly synergistic—high yield value, high viscosity, good electrolyte compatibility, and good high-temperature viscosity stability.
  - Carrageenans are synergistic in viscosity and yield value, thixotropic.
  - Tragacanth is synergistic in viscosity and yield value, thixotropic.
  - Polyacrylates are strongly synergistic in viscosity and yield value (form rigid gels and thick pourable systems).
  - Carbomers generally demonstrate limited compatibility, and tend to cause flocculation.

REFERENCES

[1] Product literature. Covington, KY: Ashland Inc., www.ashland.com.
[2] Lochhead R. History of polymers in hair care (1940-Present). Cosm Toil 1988; 103(12):23–61.

[3] Preservative directory. www.thefreelibrary.com.
[4] Preservative directory. www.Happi.com.
[5] US pharmacopeia (USP<51>).
[6] Bartulín J, Zunza H, Parra ML, Rivas BL. Chelating polymers. Polym Bull October 16, 1986;16(4):293–8.
[7] Wiechers JW, Kelly CL, Blease TG, Dederen JC. Formulating for efficacy. IFSCC Mag 2004;7(1):13–20.
[8] Barry BW. Dermatological formulations. New York, USA: Marcel Dekker; 1983. p. 128.
[9] Heena C, Rana AC, Saini S, Singh G. Effect of chemical penetration enhancers on skin permeation: a review. Int Res J Pharm 2011;2(12):120–3.
[10] Lopez-Cervantes M, Marquez-Mejia E, Cazares-Delgadillo J, Quintanar-Guerrero D, Ganem-Quintanar A, Angeles-Anguiano E. Chemical enhancers for the absorption of substances through the skin: laurocapram and its derivatives. Drug Dev Ind Pharm 2006;32:267–86.
[11] Encyclopedia of polymers and thickeners. Cosm Toil 2002; 117(12):61–120.
[12] Fox C. Sticks and gels – patent and literature update. Cosm Toil 1987;102(10):33–63.
[13] Williams PW, Phillips GO. Handbook of hydrocolloids, vol. 2. Cambridge: Woodhead; 2000. p. 28.
[14] Product literature. Philadelphia, PA: FMC Corp., www.fmc.com.
[15] Chaplin M. Water structure and behavior: guar gum. London South Bank University; April 2012.
[16] Smolinske SC. Handbook of food, drug, and cosmetic excipients. 1992. p. 7.
[17] Product literature. Minneapolis, MN: Cargill Inc., www.cargill.com.
[18] Barrére GC, Barber CE, Daniels MJ. Int J Biol Macromol 1986;8(6):372–4.
[19] Garcia-Ochoa F, Santos VE, Casas JA, Gomez E. Xanthan gum: production, recovery, and properties. Biotechnol Adv 2000;18:549–79.
[20] Crawford RL. Lignin biodegradation and transformation. New York: John Wiley and Sons; 1981.
[21] Howe AM, Flowers AE. Allured's Cosmet Toilet Mag 2000;115(12):63–9.
[22] Product literature. The Dow Chemical Company, www.dow.com.
[23] Product literature. Wickliffe, OH: The Lubrizol Corp., www.lubrizol.com.
[24] Product literature. BASF Corp., www.basf.com.
[25] Karmarkar AB. The use of polymers in pharmaceutical, biomedical and engineering fields. Pharmainfo.net. 2008. 10/27.
[26] Bohorquez M, Koch C, Trygstad T, Pandit N. A study of the temperature-dependent micellization of pluronic F127. J Colloid Interface Sci 1999;216:571–82.
[27] Cabana A, Ait-Kadi A, Juhasz J. Study of gelation process of polyethylene oxide-polypropylene oxide-polyethylene oxide copolymer (poloxmer 407) aqueous solutions. J Colloid Interface Sci 1997;190:307–12.
[28] Henry RL, Schmolka IR. Burn wound coverings and the use of poloxamers preparations. Crit Rev Biocompat 1989;5:207–20.
[29] Haaf F, Sanner A, Straub F. Polymers of N-vinylpyrrolidone: synthesis, characterization and uses. Polym J 1985;17(1):143–52.
[30] Folttmann H, Quadir A. Polyvinylpyrrolidone (PVP) – one of the most widely used excipients in pharmaceuticals: an overview. Excipient update. Drug Deliv Technol June 2008;8(6):22–7.
[31] Abu-Jdayil B. Rheology of sodium and calcium bentonite–water dispersions: effect of electrolytes and aging time. Int J Min Process 2011;98(3).
[32] Minerals technical data. Norwalk, CT: R T Vanderbilt Holding Co. Inc., www.rtvanderbilt.com.

CHAPTER 6

Aerosols and Nasal Sprays

Contents

Many types of aerosol products exist—such as deodorants, antiperspirants, hair sprays, insecticides, oils, polishes, and paints. The pressurized metered-dose inhaler (pMDI) is one type of aerosol product that is used to deliver drugs as a mist or spray via the pulmonary system. The common formulation types used for aerosol and nasal spray products are solutions, suspensions, and emulsions. This chapter looks at the manufacturing processes for aerosol and nasal spray products, the types of excipients that are used in aerosol and nasal spray formulations, and the effect that the excipients can have on the physical characteristics of the expelled product.

6.1 AEROSOLS

Aerosols are complex delivery systems comprising the concentrate, the container, a valve/actuator, and the propellant. Parameters that can affect the physical characteristics of the aerosol (e.g., droplet size, foam structure)

Essential Chemistry for Formulators of Semisolid and Liquid Dosages
http://dx.doi.org/10.1016/B978-0-12-801024-2.00006-6

include the formulation ingredients, the viscosity and surface tension of the formulation, the design of the valve, and the dimensions and shape of the actuator orifice. When formulating an aerosol product, the entire package must be taken into consideration.

Both homogeneous and heterogeneous systems can be formulated when liquefied gases are used as propellants [1]. A homogeneous (two-phase) system is formed when the product concentrate is dissolved or dispersed in a liquefied propellant (or propellant–solvent mixture). The propellant exists in both the liquefied phase and the vapor phase. When the aerosol valve is actuated, some liquefied propellant (or propellant–solvent mixture) containing the product concentrate is expelled from the container. These aerosols are designed to produce a fine mist or wet spray. The two-phase system is commonly used to formulate aerosols for inhalation or nasal application.

A heterogeneous (three-phase) system consists of a water-immiscible liquid propellant, a propellant-immiscible liquid (usually water) that contains the product concentrate, and the vapor phase. When the aerosol valve is actuated, the pressure of the vapor phase causes the liquid phase to rise in the dip tube and be expelled from the container. When some of the liquefied propellant is mixed into the expelled aqueous phase, the evaporating propellant produces bubbles in the formulation, resulting in a foam product. With this type of formulation, the container should be shaken immediately prior to use.

Topical aerosols typically contain 50–90% propellant, and oral and nasal aerosols up to 99.5% propellant. As the percentage of propellant increases, so do the degree of dispersion and the fineness of the spray. As the percentage of propellant decreases, the wetness of the spray will increase. The droplet sizes of topical aerosol sprays can vary from 50 to 100 μm.

In the case of pMDIs, the size and shape of the particles are critical to their deposition in the lungs [2]. The size is defined by what is called the mass median aerodynamic diameter (MMAD)—that is, the diameter of a particle in which 50% of the aerosol mass is greater and the other 50% is smaller. It is generally considered that particles with an MMAD greater than 10 μm are deposited in the oropharynx, those measuring between 5 and 10 μm in the central airways, and those from 0.5 to 5 μm in the small airways and alveoli. Therefore, for topical respiratory treatment, the optimal particle size is an MMAD between 0.5 and 5 μm.

6.1.1 Types of Formulation

An aerosol formulation consists of two components, the product concentrate and the propellant. The product concentrate is the active drug combined with additional ingredients or cosolvents required to make a stable and efficacious product. The product concentrate can be water or solvent based [3], and take the form of a solution, a suspension, or a semisolid (e.g., emulsion) [1]. The propellant provides the force that expels the product concentrate from the container, and is responsible for the delivery of the formulation in the proper form (i.e., spray, foam). Some propellants can also act as the solvent or vehicle for the product concentrate.

- Water based: water-based products are normally filled in phenolic-, urethane-, or epoxy-lined cans. When they are filled into unlined cans, an effective corrosion inhibitor is needed within the formulation. The propellants commonly used for aqueous products are hydrocarbon, hydrocarbon blends, dimethyl ether, hydrofluorocarbons (HFC), and nitrogen. The only propellant that is not normally used with aqueous formulations is carbon dioxide. In an aqueous medium, this propellant will form carbonic acid—causing potential container corrosion problems and pH-related product stability issues. Typical ingredients used in aqueous formulations include corrosion inhibitors, antioxidants, and biocides.

- Solvent based: unlined containers are normally used for solvent-based formulations, because most solvents will interact with or dissolve the internal lacquer of the can. If the moisture content of the concentrate is greater than 0.1%, a corrosion inhibitor may be necessary. Propellants typically used for solvent-based formulations are hydrocarbons, hydrocarbon blends, HFC, dimethyl ether, carbon dioxide, and nitrogen.

- Solution aerosols: two-phase systems consisting of the product concentrate in a propellant, a mixture of propellants, or a mixture of propellant and solvent. As many propellant or propellant–solvent mixtures are nonpolar, they tend to be poor solvents for the product concentrate. Ethanol is the most commonly used solvent. Propylene glycol, dipropylene glycol, ethyl acetate, hexylene glycol, and acetone have also been used.

- Suspension aerosols: these consist of a product concentrate that is insoluble in the propellant or propellant–solvent mixture, or when the use of a cosolvent is not desirable. Antiasthmatic drugs, steroids, and antibiotics are examples of suspension aerosol products. When the valve is actuated, the suspension formulation is expelled as an aerosol and the propellant rapidly vaporizes leaving behind a fine dispersion of the product concentrate.

Formulation issues encountered with suspension aerosols include (1) controlling the size of the dispersed aerosolized particles; (2) agglomeration and particle size growth leading to valve clogging; and (3) the moisture content of the formulation ingredients. Lubricants and surfactants are used to overcome the difficulties of particle size agglomeration, particle growth, and the associated clogging problems.

- Emulsion aerosols: the product concentrate consists of the active ingredient, aqueous and/or nonaqueous vehicles, and a surfactant. Depending on the components, the expelled product can form a stable foam (e.g., shaving cream) or a quick-breaking foam (collapses in a relatively short time). Foams are produced when the propellant is in the internal phase of the emulsion. The surfactant forms a film at the interface between the propellant and the aqueous phase. Thick, tight surfactant layers produce structured foams that are capable of supporting their own weight.

6.1.2 Typical Aerosol Excipients

The US Food and Drug Administration (FDA) produces a database of inactive ingredients that are approved for use in different dosage forms [4]—including the maximum approved concentrations. Table 6.1 lists those approved for aerosol products (as of September 16, 2013), as well as the typical function of the excipients.

6.1.2.1 Examples of Uses and Effects of Aerosol Formulation Excipients

In the case of aqueous (three phase) aerosol formulations, the water is typically immiscible with the propellant [5]. The solubility of the propellant in water can be increased by adding a cosolvent (such as ethanol) or nonpolar surfactants (e.g., esters of oleic, palmitic, or stearic acid).

The solubility of surfactants in the propellant is an important consideration for emulsion formulations. The polarity of the surfactant, as indicated by the hydrophilic–lipophilic balance (HLB), will give an indication of this solubility in different propellant systems. A high HLB value indicates that the surfactant is highly hydrophilic, and a low HLB value indicates a high lipophilic nature. Surfactants such as sorbitan trioleate (HLB 1.8) and oleic acid (HLB 1.0) are readily soluble in chlorofluorocarbon (CFC) propellants, but have low solubility in hydrofluoroalkane propellants [6].

Surfactants used in emulsion aerosols include fatty acids, anionic, and nonionic surfactants. Nonionic surfactants present fewer compatibility problems than anionic surfactants because they carry no electronic charge.

Table 6.1 Ingredients approved for use in aerosol products

Ingredient	Dosage Form	Maximum Potency Amount (%)	Typical Function
DL-α-tocopherol acetate	Emulsion foam	0.002	Antioxidant
Alcohol	Inhalation, metered	95.89	Cosolvent, preservative
	Inhalation, spray	33	
	Oral, metered	0.0001	
	Topical emulsion foam	58.21	
Alcohol dehydrated	Inhalation, metered	34.548	Cosolvent, preservative
	Nasal	2	
	Nasal, metered	0.7	
	Oral, metered	0.4716	
	Respiratory (inhalation), metered	10	
Alcohol denatured	Topical emulsion foam	60.43	Cosolvent, preservative
Alcohol diluted	Topical	68	Cosolvent, preservative
Ammonium lauryl sulfate	Topical	68.5	Surfactant
Ammonyx	Topical, metered	3	Surfactant, emulsifier, viscosity builder
Anhydrous citric acid	Topical	0.01	pH adjustment/buffer, flavoring
	Topical emulsion foam	0.08	
Apaflurane	Oral, metered	0.0068	Propellant
	Respiratory (inhalation), metered	7.46	
Ascorbic acid	Inhalation, metered	95.95	Antioxidant
	Inhalation, spray	0.1	
Butane	Sublingual, metered	2.1998	Propellant
	Topical emulsion foam		

Continued

Table 6.1 Ingredients approved for use in aerosol products—cont'd

Ingredient	Dosage Form	Maximum Potency Amount (%)	Typical Function
Butylated hydroxytoluene	Topical	0.1	Antioxidant
	Topical emulsion foam	0.1	
Caprylic/capric triglyceride	Topical spray		Moisturizer
Caprylic/capric/succinic triglyceride	Sublingual		Moisturizer
	Sublingual, metered	1.0359	
Ceteareth-12	Topical emulsion foam	5	Surfactant, emulsifier
Cetyl alcohol	Rectal, metered	0.162	Emollient, emulsifier,
	Topical	1.16	thickener
	Topical, metered	10	
	Topical emulsion foam	3.226	
Cetylpyridinium choride	Inhalation, metered		Preservative
	Oral, metered		
Citric acid	Inhalation, metered	0.0002	pH adjustment/buffer,
	Topical	0.11	flavoring
	Topical emulsion foam	0.11	
Coco diethanolamide	Topical, metered	3	Foaming agent, emulsifier
Cyclomethicone	Topical emulsion foam	5.26	Cosolvent, humectant
Cyclomethicone 5	Topical	2.5	Cosolvent, humectant
D&C red no. 28	Topical	0.0007	Coloring
Dichlorodifluoromethane	Inhalation, metered	74.029	Propellant
	Intrapleural, metered		
	Nasal	1.72	
	Nasal, metered	13.5	
	Rectal, metered		
	Sublingual		
	Topical	8	
	Topical, metered		
	Topical spray		
	Topical emulsion foam		

Ingredient	Route	%	Function
Dichlorofluoromethane	Oral, metered	35	Propellant
Dichlorotetrafluoroethane	Inhalation, metered	51.12	Propellant
	Nasal	0.86	
	Nasal, metered	9	
	Rectal, metered		
	Sublingual		
	Topical		
	Topical, metered		
	Topical emulsion foam		
Diisopropyl adipate	Emulsion foam	5	Emollient
Edetate disodium	Topical emulsion foam	0.05	Chelator
Ether	Sublingual		Solvent
Fluorochlorohydrocarbons	Inhalation, metered		Propellant
	Inhalation spray		
Fragrance P O FL-147	Topical	0.274	Fragrance
	Topical, metered	0.1	
Fragrance RBD-9819	Topical emulsion foam	0.1	Fragrance
Gluconolactone	Topical, metered	0.25	Emollient
Glycerin	Topical	3	Cosolvent, humectant
	Topical emulsion foam	2.11	
Glyceryl stearate	Topical emulsion foam		Emollient, emulsifer
Hydrocarbon	Rectal, metered	5.21	Propellant
Hydrochloric acid	Inhalation, metered	1.72	pH adjustment
	Topical emulsion foam		
Isobutane	Topical	6	Propellant
	Topical spray		
Isopropyl alcohol	Topical, metered	4	Cosolvent
	Topical spray	10	

Continued

Table 6.1 Ingredients approved for use in aerosol products—cont'd

Ingredient	Dosage Form	Maximum Potency Amount (%)	Typical Function
Isopropyl myristate	Topical	9.3	Emollient
Lactic acid	Topical emulsion foam	7.9	
	Topical	1.05	Emollient, keratolytic
	Topical emulsion foam	1.05	
Lanolin, ethoxylated	Topical, metered	1.5	Surfactant, emulsifier
Laureth-23	Topical	0.45	Surfactant, emulsifier
	Topical emulsion foam	1.075	
Lecithin	Inhalation, metered	0.0002	Emulsifier
Lecithin, hydrogenated soy	Inhalation, metered	0.28	Emulsifier
Lecithin, soybean	Inhalation, metered	0.1	Emulsifier
Levomenthol	Sublingual, metered	0.002	Topical analgesic, flavoring
Light mineral oil	Topical emulsion foam	6	Cosolvent
Menthol	Inhalation, metered	0.0502	Topical analgesic, flavoring
	Sublingual		
Methylparaben	Rectal, metered	0.09	Preservative
	Topical emulsion foam	0.108	
Mineral oil	Topical spray	1.67	Cosolvent
Nitric acid	Inhalation, metered	7.5	pH adjustment
Norflurane	Inhalation, metered	5.451	Propellant
	Nasal	5.4234	
	Oral, metered		
	Oral solution for inhalation	91.844	
	Respiratory (inhalation), metered		
Oleic acid	Inhalation, metered	0.267	Emulsifier
	Nasal, metered	0.132	
	Oral, metered	0.0003	
	Respiratory (inhalation), metered		

Component	Route/form	Amount	Function
Parabens	Topical, metered	10	Preservative
PEG-2 stearate	Topical emulsion foam	1.5	Surfactant, emulsifier
PEG-75 lanolin	Topical		Surfactant
Peppermint oil	Sublingual	0.0222	Flavoring
	Sublingual, metered		
Petrolatum, white	Topical emulsion foam	7.9	Cosolvent
Phenoxyethanol	Topical	1.05	Preservative
	Topical emulsion foam	1.05	
Polyethylene glycol 1000	Respiratory (inhalation), metered	0.0224	Lubricant
	Topical emulsion foam		
Polyethylene glycol 400	Oral, metered	5.5	Lubricant
Polyoxyl 20 cetostearyl ether	Topical	4.74	Emulsifier
Polyoxyl 40 stearate	Topical emulsion foam	1.075	Surfactant, emulsifier
Polysorbate 60	Topical emulsion foam	0.42	Surfactant, emulsifier
	Topical	0.42	
Potassium citrate	Topical	0.26	pH adjustment/buffer
	Topical emulsion foam	0.17	
Potassium citrate anhydrous	Topical	0.03	pH adjustment/buffer
Potassium hydroxide	Topical emulsion foam		pH adjustment
Potassium sorbate	Topical emulsion foam	0.2	Preservative
Povidone K25	Respiratory (inhalation), metered	0.0001	Suspending agent
Propane	Topical spray		Propellant
	Topical emulsion foam		
Propellant A-46	Topical emulsion foam		Propellant

Continued

Table 6.1 Ingredients approved for use in aerosol products—cont'd

Ingredient	Dosage Form	Maximum Potency Amount (%)	Typical Function
Propylene glycol	Rectal, metered	18	Cosolvent, preservative
	Topical	19.2	
	Topical, metered	21.05	
	Topical emulsion foam		
Propylene glycol/diazolidinyl urea/methylparaben/ propylparaben	Topical, metered	12.5	Preservative
Propylparaben	Rectal, metered	0.009	Preservative
	Topical emulsion foam	0.011	
Quaternium-52	Topical emulsion foam		Surfactant
Saccharin	Inhalation, metered	0.1127	Flavoring
	Oral, metered		
Saccharin sodium	Inhalation, metered	0.045	Flavoring
SD alcohol 40	Topical emulsion foam	46	Cosolvent, preservative
SD alcohol 40-2	Topical	57.65	Cosolvent, preservative
SD alcohol 40B	Topical	56.09	Cosolvent, preservative
	Topical emulsion foam	56.09	
Sodium chloride	Respiratory (inhalation) solution	0.225	Tonicity
Sodium hydroxide	Inhalation, metered		pH adjustment
	Respiratory (inhalation) solution		
Sodium phosphate dibasic, anhydrous	Topical emulsion foam	0.09	pH adjustment/buffer
Sorbic acid	Topical emulsion foam	0.15	Preservative
Sorbitan monolaurate	Topical	4.74	Surfactant, emulsifier
	Topical emulsion foam	4.74	

Substance	Route	Value	Function
Sorbitan trioleate	Inhalation, metered	0.0694	Emulsifier
	Nasal, metered	0.0175	
Steareth-10	Rectal, metered	0.225	Surfactant, emulsifier
	Topical, metered	6.25	
	Topical emulsion foam		
Steareth-40	Rectal, metered	1.35	Surfactant
	Topical emulsion foam	0.92	
Stearyl alcohol	Topical	0.53	Emollient, emulsifier, thickener
	Topical emulsion foam	1	
Sulfacetamide sodium	Topical emulsion foam	3.013	Topical antibiotic
Sulfuric acid	Respiratory (inhalation) solution		pH adjustment
Trichloromonofluoromethane	Inhalation, metered	33.831	Propellant
	Nasal, metered	0.9	
	Oral, metered	65	
	Topical		
	Topical spray		
Trideceth-10	Topical, metered	4	Surfactant, emulsifier
Trolamine	Rectal, metered	1	Emulsifier, pH adjustment
	Topical, metered		
	Topical emulsion foam		
Wax, emulsifying	Rectal, metered	1.5	Emulsifier
	Topical, metered	20	

With two-phase (solution) aerosol systems, the addition of a cosolvent to the formulation (such as ethanol or glycerin) will produce larger aerosol droplets [5].

Lubricants such as isopropyl myristate and light mineral oil, and surfactants such as sorbitan trioleate, oleic acid, and lecithin have been used to control particle size agglomeration and growth, which in turn can lead to aerosol valve clogging problems.

The active ingredients in pMDI are dissolved or suspended in a mixture which may include propellants, volatile and nonvolatile cosolvents, surfactants, polymers, suspension stabilizers, and bulking agents [6]. The propellant composes the bulk of the formulation, and is the driving force to deliver droplets containing the drug and excipients to the lungs by atomizing the formulation. The primary cosolvent used in pMDI formulations is ethanol. As the concentration of ethanol is increased, the vapor pressure of the formulation decreases and the droplet size increases. These larger droplets tend to deposit in the mouth and throat, thereby decreasing the dosing efficiency.

With suspension pMDI formulations, variability in the delivered dose can be caused by drug deposition on the aerosol canister components, or flocculation and creaming within the formulation. Submicron bulking agents have been used to improve dosing reproducibility by forming a loosely flocculated drug–excipient matrix that minimizes the ability of the drug to migrate out of the formulation. Examples of bulking agents that have been used are saccharides (e.g., lactose and maltose), amino acids (e.g., glycine and leucine), and salts (e.g., sodium chloride).

6.1.3 Propellants

Propellants commonly used in pharmaceutical aerosols cover both liquefied gases (CFC, hydrocarbons, hydrochlorofluorocarbons (HCFC), and HFC) and compressed gases (nitrogen, nitrous oxide, and carbon dioxide) [1].

When liquefied-gas propellants are sealed in an aerosol container with the product concentrate, an equilibrium is established between the propellant that remains liquefied and the portion that vaporizes and occupies the head space of the container. The vapor pressure at equilibrium is characteristic for each propellant at a given temperature, and is independent of the quantity of liquefied phase present. When the aerosol canister valve is actuated, the pressure forces the liquid phase up the dip tube and out of the container. When the propellant reaches the air, it evaporates due to the drop in pressure and leaves the product concentrate as airborne liquid droplets or dry particles. As the liquid (concentrate plus propellant) is expelled from the container, the equilibrium between the liquefied and vapor phases of the propellant is rapidly reestablished. This evaporation of the liquefied propellant causes the product to cool (due to the latent heat of vaporization of the liquid propellant), resulting in a very slight pressure drop within the container. As the pressure drop is small, the product is continuously released at an even rate and with the same propulsion. The product temperature (and vapor pressure) quickly returns to normal condition once the valve is released.

An important characteristic of any aerosol is the density of the propellant system. Chlorofluorocarbons, HCFC, and HFC are denser than water, and will reside on the bottom of the container. Hydrocarbons are less dense than water, and will reside on top of the aqueous layer. The length of the dip tube must take into account where the concentrate and propellant reside within the can—for example, with hydrocarbon propellants, the dip tube can extend through the liquid propellant to the bottom of the container.

For pMDI formulations, the propellants should be toxicologically safe, nonflammable, and chemically inert. They are usually liquefied gases, and should provide the same vapor pressure regardless of whether the pMDI canister is full or nearly empty, to provide the same delivered dose throughout the lifetime of the inhaler [6].

Foam aerosols are a three-phase system in which the liquid propellant is emulsified with the product concentrate. When the valve is actuated, the emulsion is dispensed and the entrapped propellant evaporates forming bubbles within the formulation, producing foam. To facilitate the formation of foam, some aerosols are shaken prior to use to better disperse the propellant throughout the product concentrate. If a dip tube is present, the container is used while being held upright. If there is no dip tube, the container must be inverted prior to use so that the liquid phase is in direct contact with the valve. Propellants used with foam aerosols include propane/isobutane blends, isobutane, and the less flammable difluoroethane, as well as compressed gases (nitrous oxide and carbon dioxide).

Aerosols using compressed gases as propellants operate essentially as a pressure package. The propellant will usually only be in the head space, and the pressure of the gas forces the product concentrate out of the container in essentially the same form as it was placed into the container. The pressure in the container decreases as the product concentrate is expelled, and the gas expands to occupy the newly vacated space. As a result, the initial pressure used with compressed gases is higher than that needed for liquefied-gas aerosols.

6.1.3.1 Numerical Designations for Fluorinated Hydrocarbon Propellants

The numerical designation for fluorinated hydrocarbon propellants has been designed such that the chemical structure of the compound can be determined from the number.

- The digit at the extreme right = the number of fluorine atoms
- The second digit from the right = (the number of hydrogen atoms) plus 1
- The third digit from the right = (the number of carbon atoms) minus 1. If this third digit is "0", it is omitted and a two-digit number is used
- The capital letter "C" is used before a number to indicate the cyclic nature of a compound
- The small letters following a number are used to indicate decreasing symmetry of isomeric compounds. The most symmetrical compound is given the designated number, and all other isomers are assigned a letter (i.e., a, b, etc.) in the descending order of symmetry
- The number of chlorine atoms in a molecule may be determined by subtracting the total number of hydrogen and fluorine atoms from the total number of atoms required to saturate the compound

6.1.3.2 Chlorofluorocarbon (CFC) Propellants

Examples of CFC propellants are trichlorofluoromethane (CCl_3F, Propellant 11); dichlorodifluoromethane (CCl_2F_2; Propellant 12); and 1,2-dichloro-1,1,2,2-tetrafluoroethane ($ClF_2C-CClF_2$, Propellant 114). These propellants are gases at room temperature that can be liquefied by cooling below their boiling point or by compressing at room temperature. These liquefied gases also have a very large expansion ratio compared to the compressed gases (e.g., nitrogen, carbon dioxide).

For many years, CFC propellants were widely used in aerosol products. However, due to their role in depleting the ozone layer of the atmosphere, their phase out and substitution with more environmentally friendly propellants started in the late 1980s under the Montreal Protocol. The US Environmental Protection Agency produces a list of acceptable propellant substitutes based on factors such as ozone depletion potential, global warming potential, toxicity, flammability, and exposure potential [7]. The use of CFC propellants has now been restricted to aerosols used in the treatment of asthma and chronic obstructive pulmonary disease. Propellants 11, 12, and 114 are the choice for oral, nasal, and inhalation aerosols due to their relatively low toxicity and inflammability.

6.1.3.3 Hydrochlorofluorocarbon (HCFC) and Hydrofluorocarbon (HFC) Propellants

HCFC and HFC propellants break down in the atmosphere at a faster rate than the CFCs, resulting in a lower ozone-depleting effect.

Examples of HCFC and HFC propellants are chlorodifluoromethane ($CHClF_2$; Propellant 22); 1,1,1,2-tetrafluoroethane (CF_3CH_2F; Propellant 134a); 1-chloro-1,1-difluoroethane (CH_3CClF_2; Propellant 142b); 1,1-difluoroethane (CH_3CHF_2; Propellant 152a); and 1,1,1,2,3,3,3-heptafluoropropane (CF_3CHFCF_3; Propellant 227).

Propellants 22, 142b, and 152a are used in topical pharmaceuticals. These three propellants have good miscibility with water, which makes them more useful as solvents compared to some other propellants. Medicinal aerosols such as asthma inhalers use the HFC propellants 134a, 227, or a combination of the two.

6.1.3.4 Hydrocarbon Propellants

Hydrocarbon propellants are used in topical pharmaceutical aerosols because of their environmental acceptance, low toxicity, and lack of reactivity. They are useful in three-phase (two-layer) aerosol systems because they are immiscible with water and have a density less than 1. The hydrocarbons remain on top of the aqueous layer, and provide the force to expel the contents from the container. As they contain no halogens, hydrolysis does not occur making them good propellants for water-based aerosols. The main drawback is that they are flammable, and can explode.

Propane (C_3H_8; Propellant A-108), butane (C_4H_{10}; Propellant A-17), and isobutane (C_4H_{10}; Propellant A-31) are the most commonly used hydrocarbons. They are used alone, as mixtures, or mixed with other liquefied gases to obtain the desired vapor pressure, density, and degree of flammability.

6.1.3.5 Compressed Gases

In the case of compressed-gas propellants, the pressure of the compressed gas in the headspace of the aerosol container expels the product concentrate in essentially the same form as it was placed into the container. Unlike aerosols containing liquefied-gas propellants, there is no propellant reservoir. As a result, higher gas pressures are required for aerosols that use compressed gases, and the pressure within the aerosol diminishes as the product is used.

Gases such as nitrogen (N_2), nitrous oxide (N_2O), and carbon dioxide (CO_2) have been used as aerosol propellants for products that are dispensed as fine mists, foams, or semisolids—including food products, dental creams, hair preparations, and ointments. Due to their low expansion ratio, the sprays are fairly wet, and the foams are not as stable as those produced by liquefied-gas propellants.

6.1.4 Particle Engineering and pMDIs

For pMDI suspension formulations, the size of the crystalline active ingredient needs to be controlled (or reduced) to be suitable for inhalation [6]. This can be achieved using mechanical particle size reduction techniques such as ball or jet milling. The drawbacks to these mechanical techniques are that they can affect the crystallinity of the material, and they may produce nonspherical particles (the presence of flat surfaces could increase the adhesion between micronized particles). Particle engineering technologies [8] including spray drying, wet polishing, and controlled crystallization may be used as alternatives to the mechanical techniques.

One of the major concerns for suspension pMDI formulations is the uniformity of the expelled dose. Phase separation, flocculation, agglomeration, and interaction of the drug particles with other formulation ingredients or the container are all possible causes of nonuniform formulations. Surfactants have been used to combat these problems and stabilize the dispersion by decreasing the electrostatic attraction between the micronized drug particles. Other approaches have included using bulking agents and phospholipid encapsulation. The drug particles must then be readily and homogeneously resuspended in the formulation by shaking prior to use.

In the case of solution pMDIs, surfactants are used to increase the solubility of the drug particles and to overcome valve-sticking issues.

6.1.5 Manufacturing Process

Two typical processes are used for filling aerosol products—cold filling and pressure filling [9]:
* Cold filling

The formulation concentrate consists of a solution (or suspension) of the functional ingredient(s) dissolved in a solvent (or suspended in a carrier) that is a liquid at room temperature. In the case of drug products and pMDIs, the functional ingredient would be the active pharmaceutical ingredient (API). The bulk propellant (which forms the rest of the formulation) is placed into a prechilled mixing vessel at a temperature low enough to ensure that the propellant is also in liquid form. The concentrate is added to the cold propellant in the mixing vessel, and the entire formulation is mixed to ensure homogeneity.

The chilled-liquid formulation is then dispensed into the open aerosol canisters. The heavy vapors of the cold-liquid propellant will generally displace the air present within the canister. A valve is placed on top of each canister and crimped into place, forming a seal between the top of the canister and a rubber gasket within the valve. Each completed aerosol is checked for weight to ensure the correct amount of formulation is in the container.

Filled containers may then be passed through a heated water bath to ensure a proper seal has been formed and that there are no gaps through which the propellant may leak. The hot water bath also serves to warm the aerosol to room temperature. The formulation in the canister remains a liquid, due to the fact that it is under pressure.

As the concentrate and propellant are mixed in bulk prior to being dispensed into the aerosol canister, the cold-filling process is well suited to both solution and suspension formulations—including suspensions that contain a high loading of solid material. Additionally, the chilling stage of the cold-filling process can control the crystallization of the product and therefore the particle size of the API.

• Pressure filling

In contrast to cold filling, the pressure-filling process uses pressure instead of low temperature to condense the propellant. The liquid concentrate is made in the same way as for cold filling, but the propellant is held in a pressurized vessel in liquid form. There are two different methods for pressure filling the aerosol canister:

1. two-stage pressure filling: the concentrate is dispensed into an open aerosol canister. A valve is then placed on top of the canister and crimped into position to form the seal. The propellant is driven under pressure through the valve and into the canister. Using this method, the mixing of the concentrate and propellant occurs in the canister, rather than in a bulk formulation tank

2. single-stage pressure filling: the concentrate and propellant are combined in a mixing vessel and held under pressure. An empty aerosol canister is assembled with a valve which is then crimped into place. The complete formulation is driven under pressure through the valve into the canister

As with the cold–filling process, the filled aerosols are checked for weight and leakage. Functionality testing of the valves at this stage also serves to rid the dip tube of pure propellant prior to consumer use.

Any entrapped air in the pressure-filled package might be ignored if it does not interfere with the stability of the product, or it may be evacuated prior to or during the filling process.

Pressure filling is used for most pharmaceutical aerosols. It has the following advantages: there is less danger of moisture contamination of the product (caused by condensation occurring at the low temperatures used in the cold-filling process) and also less propellant is lost in the process.

Single–stage pressure filling is ideal for solutions due to the fact that the formulation can be readily driven back through the valve into the canister. Single–stage pressure filling can also be appropriate for suspensions that have a low loading of solid material.

Two–stage pressure filling is generally used with suspension formulations in which the loading of the solid material makes the formulation too thick for it to be dispensed into the can through the valve with repeatable accuracy.

6.2 NASAL SPRAYS

Nasal spray systems comprise the formulation, the container, and an actuator. Similar to aerosols, nasal sprays rely on the atomization of the formulation to form a plume of droplets as it is emitted from the device. Nasal sprays differ from aerosols in that manual depression of the device actuator is used to dispense the product instead of the pressure created by a propellant. The lack of a propellant simplifies the formulation to a certain extent, but adds a degree of variability to the expelled plume resulting from patient-to-patient actuation differences (force, acceleration, and velocity of actuation). Parameters that can affect the physical characteristics of the plume (e.g., droplet size, spray pattern) include the formulation ingredients, the viscosity and surface tension of the formulation, the design, dimensions, and shape of the actuator, and the technique used to actuate the nasal device. When formulating nasal spray products, both the formulation and the device must be taken into consideration [10].

The nasal route of delivery has been used for drugs that act locally (e.g., antihistamines) as well as systemically (e.g., breakthrough pain management).

Systemically, the nasal route offers the benefit that it bypasses the blood–brain barrier, delivering drugs directly to the central nervous system [11]. This should enable lower doses of active drugs to be used to achieve the same therapeutic effect when compared to other routes of administration (e.g., oral or intravenous), thereby reducing the possible side effects.

Some of the challenges with nasal delivery are generating plumes that consist of the correct size droplets to be deposited in the nasal cavity, delivering a therapeutically effective dose, and keeping the drug in place long enough for it to be effective. As for the droplet size, droplets smaller than 10 μm may travel through the nasal cavity and be deposited in the lung; and droplets larger than 300 μm might drip back out of the nose [12]. The delivered volume for nasal sprays typically lies within the 25–100 μL range. Volumes larger than 100 μL might "flood" the nasal cavity and lead to a portion of the dose either dripping back out of the nose or being swallowed. The concentration of the active drug in the formulation must be such that a therapeutically effective dose can be delivered from, at most, two 100 μL sprays—one in each nostril. To retain the drug within the nasal cavity, mucoadhesives are used within the formulation. However, as the nasal cavity "cleans" itself through a process of mucociliary clearance every 20 min [13], residence in the nasal cavity for "slow acting" drugs can still be a challenge, and may necessitate the inclusion of penetration enhancers.

6.2.1 Types of Formulation

Nasal spray formulations may be aqueous, hydroalcoholic, or nonaqueous-based solution, suspension, or emulsion systems. Depending on the type of system, the formulation will include a range of functional excipients, including solvents and cosolvents; mucoadhesive agents; pH buffers; antioxidants; preservatives; osmolality and tonicity agents; penetration enhancers; suspending agents; and surfactants. The choice of formulation type and the excipients selected will be driven by the solubility and stability of the active drug, as well as the concentration needed to deliver an efficacious dose in a typical 100 μl spray.

6.2.2 Typical Nasal Spray Excipients

The excipients approved for use in nasal spray products by the US FDA (as of September 16, 2013) are shown in Table 6.2 [4]. (Dosage forms termed "nasal aerosols" are covered in Table 6.1.)

Table 6.2 Ingredients approved for use in nasal sprays

Ingredient	Dosage Form	Maximum Potency Amount (%)	Typical Function
Acetic acid	Spray, metered		pH adjustment
Alcohol dehydrated	Spray, metered		Cosolvent
Allyl-α-ionone	Spray, metered	1	Flavoring
Anhydrous dextrose	Spray	0.275	Sweetener,
	Spray, metered	0.5	tonicity
Anhydrous trisodium citrate	Spray, metered	0.0006	pH buffer
Benzalkonium chloride	Spray solution	0.025	Preservative,
	Spray	0.119	surfactant
	Spray, metered	12.5	
Benzethonium chloride	Spray, metered	0.02	Preservative
Benzyl alcohol	Spray, metered	0.045	Preservative
Butylated hydroxyanisole	Spray, metered	0.0002	Antioxidant
Butylated hydroxytoluene	Spray, metered	0.01	Antioxidant
Caffeine	Spray, metered		Analgesic adjuvant
Carbon dioxide	Spray, metered		pH adjuster, exclude oxygen (oxidation)
Carboxymethylcel-lulose sodium	Spray, metered	0.15	Thickener, mucoadhesive
Cellulose microcrystalline/ carboxymethyl-cellulose sodium	Spray	2	Suspending agent
	Spray, metered	1.5	
Cellulose microcrystalline	Spray, metered	0.15	Suspending agent
Chlorobutanol	Spray	0.5	Preservative
	Spray, metered		
Citric acid	Spray solution		pH adjustment/ buffer
	Spray	0.26	
	Spray, metered	3.5	
Citric acid monohydrate	Spray, metered	0.42	pH adjustment/ buffer
Dextrose	Spray	5	Sweetener
	Spray, metered	5	

Table 6.2 Ingredients approved for use in nasal sprays—cont'd

Ingredient	Dosage Form	Maximum Potency Amount (%)	Typical Function
Edetate disodium	Spray solution	0.05	Chelator,
	Spray	0.1	preservative
	Spray, metered	50	enhancer
Glycerin	Spray	2.3	Cosolvent,
	Spray, metered	0.223	humectant
Hydrochloric acid	Spray solution		pH adjustment
	Spray		
	Spray, metered		
Hypromellose 2910	Spray solution	0.1	Thickener,
	Spray		suspending
	Spray, metered		agent, mucoadhesive
Hypromelloses	Spray, metered		Thickener, suspending agent, mucoadhesive
Mannitol	Spray solution	4.15	Sweetener, osmolality
Methylparaben	Spray, metered	0.7	Preservative
Nitrogen	Spray, metered		Exclude oxygen (oxidation)
Norflurane	Spray, metered		Propellant
Pectin	Spray solution	1	Thickener, mucoadhesive
Phenylethyl alcohol	Spray solution	0.5	Preservative
	Spray	0.25	
	Spray, metered	0.254	
Polyethylene glycol 3350	Spray, metered	1.5	Humectant, lubricant, surfactant
Polyethylene glycol 400	Spray, metered	20	Cosolvent, surfactant
Polyoxyl 400 stearate	Spray, metered	15	Emulsifier, surfactant, cosolvent
Polysorbate 20	Spray, metered	2.5	Emulsifier, surfactant
Polysorbate 80	Spray	0.005	Emulsifier, surfactant
	Spray, metered	10	
Potassium phosphate monobasic	Spray	0.14	pH buffer

Continued

Table 6.2 Ingredients approved for use in nasal sprays—cont'd

Ingredient	Dosage Form	Maximum Potency Amount (%)	Typical Function
Potassium sorbate	Spray, metered	0.0084	Preservative
Propylene glycol	Spray, metered	20	Humectant, cosolvent
Propylparaben	Spray solution	0.02	Preservative
	Spray, metered	0.3	
Sodium chloride	Spray	0.9	Tonicity,
	Spray, metered	1.9	osmolality
Sodium citrate	Spray	0.44	pH buffer,
	Spray, metered	68	alkalizer
Sodium hydroxide	Spray solution		pH adjustment
	Spray		
	Spray, metered	0.004	
Sodium phosphate dibasic	Spray		pH buffer
Sodium phosphate dibasic, anhydrous	Spray	0.011	pH buffer
Sodium phosphate dibasic, dihydrate	Spray, metered	0.3	pH buffer
Sodium phosphate dibasic, dodecahydrate	Spray, metered	14.3	pH buffer
Sodium phosphate dibasic, heptahydrate	Spray solution		pH buffer
	Spray, metered	0.486	
Sodium phosphate monobasic, anhydrous	Spray, metered	0.019	pH buffer
Sodium phosphate monobasic, dihydrate	Spray, metered	4.2	pH buffer
Sorbitol	Spray	6.4	Sweetener,
	Spray, metered	2.86	osmolality
Sorbitol solution	Spray solution		Sweetener,
	Spray, metered		osmolality
Sucralose	Spray solution	0.15	Sweetener
	Spray	0.15	
	Spray, metered		
Sulfuric acid	Spray	0.4	pH adjustment
Trisodium citrate dihydrate	Spray solution	0.068	pH buffer
	Spray	0.068	

6.2.2.1 Examples of Uses and Effects of Nasal Spray Formulation Excipients

Although a few patents have been reported for nonaqueous nasal spray formulations containing poorly water-soluble drugs [14,15], the majority of nasal spray formulations are aqueous-based solutions, suspensions, or thin emulsions. Typically, aqueous-based formulations aerosolize better than nonaqueous and form a wider plume. Nonaqueous formulations have a tendency to produce a relatively narrow plume, which resembles a jet or stream [10]. Also with nonaqueous formulations, excipients such as propylene glycol and PEG 400 (which are used as solvents) may cause irritation and hyperosmolarity [16].

Viscosity modifiers are used in nasal spray formulations to reduce the rate of settling of suspended active ingredients, thereby improving the stability profile of the product. Viscosity modifiers can also act as mucoadhesives, increasing the residence time of the spray in the nasal cavity. A longer residence time within the nasal cavity will improve the chance of drug absorption through the nasal mucosa. One drawback with the use of viscosity modifiers, however, is that increasing the viscosity of the formulation will make it more difficult to aerosolize. This will affect the expelled-spray plume by decreasing the width of the plume, and increasing the size of the spray droplets [10,17].

Suspending agents are also used in nasal spray formulations to reduce the rate of settling of suspended active ingredients. Although some suspending agents will not fully dissolve in aqueous-based formulations (e.g., microcrystalline cellulose/sodium carboxymethylcellulose), their use will increase the viscosity of the formulation and affect the spray plume.

When added to solution and suspension formulations, surfactants will reduce the surface tension, making it easier to aerosolize. This should have the effect of increasing the width of the plume, and decreasing the size of the spray droplets. Adding a surfactant, however, may also have the effect of increasing the viscosity of the formulation, which would have the opposite effect on the spray characteristics. The net effect on the spray plume will depend on the sensitivity of the formulation to changes in surface tension relative to changes in viscosity. One previous study has shown that it is the increase in viscosity that dominates the effect on the spray characteristics [10].

Both the pH and tonicity of the formulation may have a significant influence on the in-use performance of a nasal spray. The local pH within the nasal cavity may affect the rate and extent of absorption of ionizable drugs [18]. An optimal pH range of 4.5–6.5 has been suggested [19]. The osmolality of the nasal spray formulation may also affect drug permeability through the nasal mucosa [20,21].

Due to the rate of mucociliary clearance, penetration enhancers have been used in an attempt to speed up drug absorption. Solvents, cosolvents, surfactants, fatty acids, and lipids are some of the chemicals that have been investigated [12].

Microbial preservatives are typically incorporated into nonsterile nasal spray formulations, as well as some multiuse sterile formulations, to control the presence and effect of microorganisms. Their use, however, can potentially cause discomfort and irritation to the nasal mucosa—especially with long-term use. Additionally, their inclusion can alter the smell and/or taste of the formulation [22]. To combat these concerns, devices have been developed that will maintain the microbial integrity of preservative-free formulations (see Section 6.2.3).

Antioxidants and chelators are examples of chemical preservatives. These would be added to a formulation in which, for example, the active drug is susceptible to oxidation, or reacts with any metal ions that may be present.

Flavors and sweeteners are added to mask any bitterness or unwanted odor in the formulation resulting from the active drug ingredient or excipients. Adding flavors and sweeteners should improve the user acceptance of the product.

6.2.3 Nasal Spray Devices

Nasal spray devices for liquid and semisolid formulations come in unit dose, bi-dose, and multidose formats. With the unit and bi-dose devices, the formulation is contained in a stoppered glass vial. When the device is actuated, a needle within the body of the actuator pierces the stopper, the stopper is depressed, and the spray is expelled through the needle. With the bi-dose device, two separate actuations are required to expel the entire product. The actuation mechanism comes to a "stop" after approximately half of the vial has been dispensed so that the other half can be dispensed into the other nostril. The vial stoppers are elastomeric (some are Teflon® coated or siliconized), and contain "ribs/fins" to create an airtight seal when inserted into the vial.

Multidose devices consist of a bottle and spray-pump actuator. A dip tube goes from the actuator into the formulation to enable it to be expelled when the actuator pump is depressed. On first use of the product, the actuator pump must be depressed several times to fill the dip tube (prime the device) before the target dose will be dispensed. The number of priming strokes and the time interval before repriming is necessary are just two of the parameters that need to be determined for a multidose nasal spray. After the dose is dispensed from a multidose device, the pump is released and the air enters the device to replace the dispensed product. Without this makeup air, a vacuum would be produced in the container, restricting subsequent dispensing of the product. With conventional actuator pumps, the makeup air comes directly from the atmosphere and could introduce microbial contamination—this is one of the reasons that preservatives are included in multidose formulations. To avoid the need to add microbial preservatives, multidose devices have been developed in which the makeup air enters the device via a 0.22- μm sterilizing filter [22].

During use, any resistance generated by the nasal spray device (e.g., the resistance of the spring in a multidose actuator) must be overcome to successfully dispense the product. The size of the dose expelled and the spray plume characteristics can be affected by patient-to-patient actuation differences (e.g., force, acceleration, and velocity of actuation). These patient-to-patient actuation differences may be derived from factors such as the physical strength of different users, the user's age, or other physical challenges the user may possess (e.g., arthritis). Devices exist that will measure the dynamic actuation parameters (actuator position, velocity, and acceleration levels) applied to the nasal device during use by volunteers [23], and studies have been performed to compare the plume characteristics of the expelled spray to in-use actuation differences [24]. These types of studies are generally performed during the development stage of the nasal spray product (to aid in device selection), and should (when possible) use a volunteer population representative of the intended user group (e.g., geriatrics).

6.2.4 Manufacturing Process

The mixing process for nasal spray formulations is similar to any other solution, suspension, or emulsion. The degree of heating and shear needed during mixing will depend on the formulation in question—for example, how easily and quickly the ingredients dissolve, or what particulate or emulsion droplet size is required. In the case of suspension formulations, the use of a recirculation loop from the mixing tank to the filling nozzles might need to be considered to prevent the suspended ingredients settling out of the formulation base either during processing or during machine stoppages.

The finished formulation is filled into either a vial (unit or bi-dose) or bottle (multidose) using dosing pumps. For the unit and bi-dose vials, because of the small fill volumes and tight filling tolerances, very accurate control of the dosing volume is needed. Unit dose vials typically have a target dose volume of 100 μL. These vials are filled with an overage of 25 μL to account for the product that gets held up in the actuator needle or remains in the vial after dispensing. With bi-dose vials, the typically fill volume range is 220–230 μL.

During the stopper insertion process for the unit and bi-dose vials, the "ribs/fins" on the stopper need to be compressed as the stopper is inserted into the vial. This allows the air to the escape from the vial, thereby preventing a buildup of pressure that could lead to the stopper being pushed back out. The stopper height is tightly controlled to minimize the headspace in the vial. Additionally, an overlay of an inert gas (e.g., N_2) might be used in the vials to help prevent oxidation of the active ingredient. Typical in-process checks are fill weight, stopper height, and container closure integrity. The stoppered vial is then assembled with the device housing and actuator components. The assembly process is controlled to ensure that the actuator components are assembled to the correct "height"—so that the needle in the actuator does not start to puncture/cut into the stopper.

The filling process for multidose bottles is similar to a standard bottle filling process. The actuators can be either a screw-on, snap-on, or crimp-on design. The application torque (for screw-on actuators) or pressure (for snap-on and crimp-on actuators) can be controlled to ensure that the device is properly sealed, without exerting too much pressure (e.g., over-torquing). As with unit and bi-dose devices, an overlay of an inert gas might be used to help prevent oxidation of the active ingredient. Additional process controls might be used with snap-on and crimp-on actuators to ensure that the device does not get partially actuated (partially primed) when the actuators are pressure fitted. Typical in-process checks for multidose devices are fill weight and container closure integrity.

REFERENCES

[1] The pharmaceutics and compounding laboratory. UNC Eshelman School of Pharmacy, www.pharmlabs.unc.edu/labs/aerosols.
[2] Tena AF, Clarà PC. Deposition of inhaled particles in the lungs. Arch Bronconeumol 2012;48(07):240–6.
[3] Geer RD. Aerosol formulation considerations. RSC Chemical Solutions; 2011. www.southernaerosol.com.
[4] Inactive ingredient database. US Food and Drug Administration, www.accessdata.fda.gov/scripts/cder/iig/index.cfm.
[5] Sunita L. An overview on: pharmaceutical aerosols. Int Res J Pharm 2012;3(9):68–75.
[6] Myrdal PB, Sheth P, Stein SW. Advances in metered dose inhaler technology: formulation development. AAPS PharmSci Tech April 2014;15(2):434–55. Epub: January 23, 2014.

[7] Substitutes for ODS in aerosol solvents and propellants. United States Environmental Protection Agency, www.epa.gov/ozone/snap/aerosol/list.html.

[8] Costa E. Particle engineering and formulation development for inhalation. Hovione Industry Webinar; May 2014. www.hovione.com.

[9] Errington R. Inhalation manufacturing: cold fill, pressure fill, and finding the right partner. ONdrugDelivery November 2012;37:30–2. www.ondrugdelivery.com.

[10] Kulkarni VS, Shaw C, Smith M, Brunotte J. Characterization of plumes of nasal spray formulations containing mucoadhesive agents sprayed from different types of device. In: Poster presentation, AAPS annual meeting. 2013.

[11] Ying W. The nose may help the brain: intranasal drug delivery for treating neurological diseases. Future Neurol 2008;3(1):1–4.

[12] Kulkarni V, Shaw C. Formulation and characterization of nasal sprays. Inhalation June 2012:10–5.

[13] Hochhaus G, Sahasranaman S, Derendorf H, Mollmann H. Intranasal glucocorticoid delivery: competition between local and systemic effects. STP Pharma Sci 2002;12:23–31.

[14] Castle J, Smith A, Cheng Y-H, Watts PJ. Non-aqueous pharmaceutical application. US patent application 2009/0233912; 2009.

[15] Hatton AG, Scott H, Hilton JE. Calcium mupirocin non-aqueous nasal spray for otis media or for recurrent acute bacterial sinusitis. US Patent 6156792; 2000.

[16] Kibbe AH, editor. Handbook of pharmaceutical excipients. 3rd ed. Am Pharm Assoc; 2000. p. 392–8. 442–444.

[17] Kulkarni V, Brunotte J, Smith M, Sorgi F. Investigating influences of various excipients of the nasal spray formulations on droplet size and spray pattern. In: Poster presentation, AAPS annual meeting. 2008.

[18] Washington N, Steele RJ, Jackson SJ, Bush D, Mason J, Gill DA, et al. Determination of baseline human nasal pH and the effect of intranasally administered buffers. Int J Pharm 2000;198:139–46.

[19] Aurora J. Development of nasal delivery systems: a review. Drug Delivery Technol 2002;2(7):70–3.

[20] Dua R, Zia H, Needham TE. The influence of tonicity and viscosity on the intranasal absorption of salmon calcitonin in rabbits. Int J Pharm 1997;147:233–42.

[21] Farina DJ. Regulatory aspects of nasal and pulmonary spray drug products. In: Kulkarni V, editor. Handbook of non-invasive drug delivery systems. Oxford, UK: Elsevier; 2010. p. 247–90.

[22] Ehrick JD, Shah SA, Shaw C, Kulkarni VS, Coowanitwong I, De S, et al. Considerations for the development of nasal dosage forms. In: Kolhe P, Shah M, Rathore N, editors. Sterile product development: formulation, process, quality and regulatory considerations Springer AAPS Press; 2013. p. 99–144.

[23] Ergo™ sensor, Marlborough, MA: Proveris Scientific Corp.

[24] Shaw CJ, Kulkarni VS, Smith M, Farina DJ, Pitluk Z, Goth W. Using quality by design approach to correlate patient usage to the in vitro performance of a nasal spray product. In: Poster presentation, Respiratory Drug Delivery. 2012.

CHAPTER 7

Preparation and Stability Testing

Contents

Liquid and semisolid formulations contain many different functional ingredients depending on their physical form and intended use—for example, pH buffers, antioxidants, microbial preservatives, chelators, osmolality agents, penetration enhancers, thickeners, emulsifiers, solvent systems, humectants, pharmaceutically active ingredients, moisturizers, fragrances, flavorings, and colorants. When and how these ingredients are added to the formulation

Essential Chemistry for Formulators of Semisolid and Liquid Dosages
http://dx.doi.org/10.1016/B978-0-12-801024-2.00007-8

will determine whether the finished formulation is successful—aesthetically, functionally, and stable. This chapter looks at how formulations are typically put together, gives examples of different mixing systems, shows how Quality by Design (QbD) can be applied to formulation development, and how to evaluate the stability of a formulation.

7.1 HOW FORMULATIONS ARE PUT TOGETHER

7.1.1 General Comments

The ideal way to put a formulation together is to keep all parts of the formulation as fluid as possible until as late in the mixing process as possible, when the formulation is thickened to its final consistency. This is true for all formulations, no matter the physical form. To do this, all the ingredients are dissolved or dispersed in their respective phase before the phases are brought together and emulsified, or gelled. This minimizes the amount of shear that needs to be used to blend ingredients into the formulation and any problems associated with shear-induced loss of viscosity.

The order of addition of the ingredients will determine when they are added during the mixing process. For example:

- If a pharmaceutically active ingredient is sensitive to pH or the presence of metal ions, the pH buffer system and chelator might need to be added before the active ingredient
- If some of the ingredients only dissolve in hot water, they are usually added to the hot water at the beginning of the process. The solution is then cooled before the other, possibly more temperature-sensitive, ingredients are added. Adding an alcohol-based fragrance to a hot solution will cause evaporative loss of the fragrance
- When carbomers are used for thickening emulsion systems and gels, the polymer is fully dispersed in the solvent system before the neutralizing base is added. On addition of the base, the polymer forms a gel structure that can be broken down by high-shear mixing, leading to a loss of formulation structure and viscosity

7.1.2 Solutions

True water-thin solutions are probably the simplest formulation to put together. The ingredients are added in a stepwise manner to the solvent system, sufficient agitation being used to ensure good dispersion of each ingredient. Each individual ingredient is usually allowed to fully dissolve before the next is added. Heat may be required to dissolve, or to speed up the dissolution, of some of the ingredients. Also, different levels of mechanical shear can be used to aid in the initial dispersion process. Examples of process systems that are used to make solution formulations are shown in Figures 7.1–7.4 and described in Section 7.2.

7.1.3 Suspensions

In the case of suspensions, ingredients that are soluble in the solvent system can be added in a similar manner to those for a solution. The insoluble ingredient that is to be suspended is added toward the end of the mixing process, prior to the suspending agent or thickener.

The suspended ingredient and suspending agent/thickener need to be uniformly dispersed in the bulk solution. These ingredients are usually powders, and can be added to the recirculating bulk via a powder eductor as shown in Figures 7.5–7.7. In addition, colloid mills or homogenizers can be added to the processing system to reduce the particle size of the suspended ingredients to the required size range. These mills/homogenizers can either be integral to the main mixing tank or on an external recirculation loop (as in Figure 7.7). As the mill/homogenizer imparts high shear to the formulation, it needs to be used prior to the swelling of the thickening/suspending agent. Once the thickening/suspending agent has swollen, high-shear pumping and mixing can result in a loss of structure and viscosity—which will adversely affect the suspending properties.

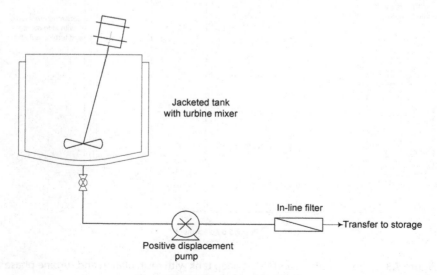

Figure 7.1 Jacketed turbine tank.

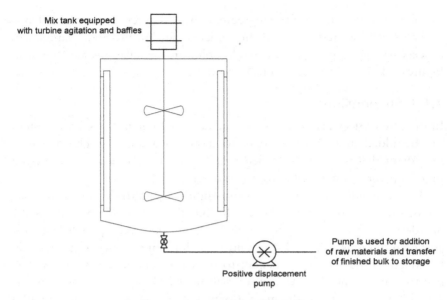

Figure 7.2 Turbine tank with baffles.

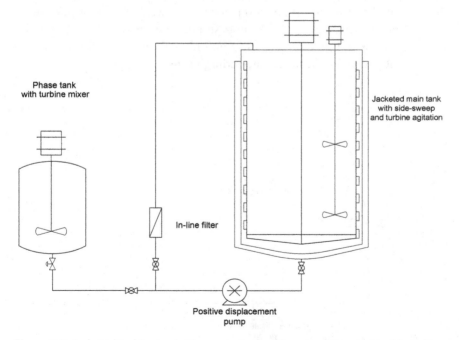

Figure 7.3 Jacketed turbine and side-sweep tank with recirculation and turbine phase tank.

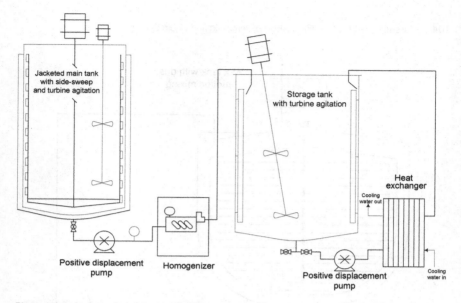

Figure 7.4 Jacketed turbine and side-sweep tank with external homogenization and stirred storage tank with heat exchanger recirculation.

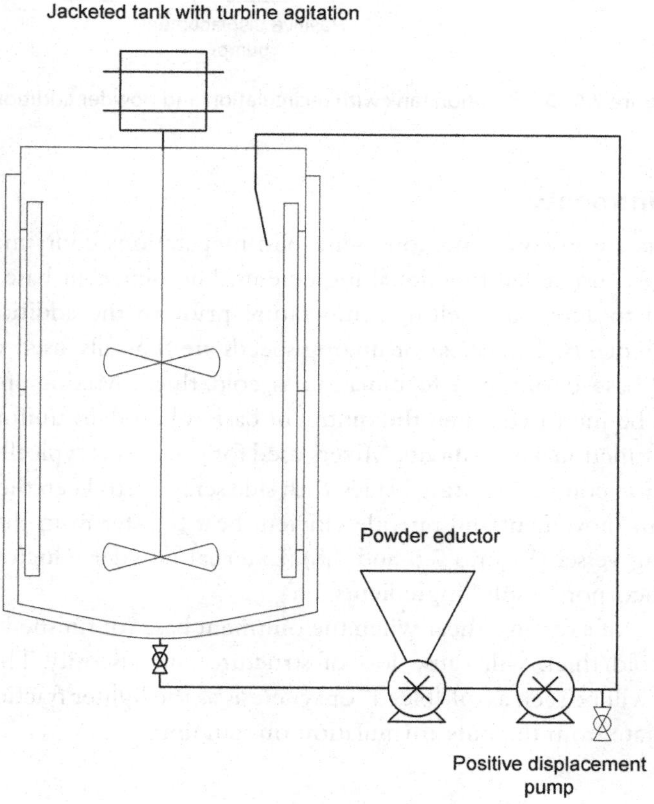

Jacketed tank with turbine agitation

Powder eductor

Positive displacement pump

Figure 7.5 Turbine tank with recirculation and powder addition.

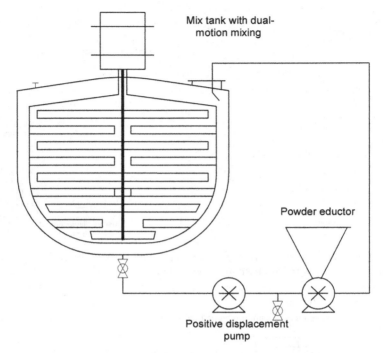

Figure 7.6 Dual-motion tank with recirculation and powder addition.

7.1.4 Ointments

Ointments are viscous, unctuous, semisolid preparations containing either dissolved or suspended functional ingredients. The ointment base needs to be heated to above its melting temperature prior to the addition of the other ingredients. Low-shear or mixing speeds are typically used when the ointment base or finished formulation is cold/thick. Mixing speeds and shear can be increased when the ointment base is liquid, to uniformly disperse the functional ingredients. Mixers used for ointments typically employ dual-motion counter-rotating blades with side scrapers, to keep the material in constant movement and provide efficient heat transfer from the walls of the mixing vessel (Figures 7.6 and 7.8). External powder eductors can be used to incorporate solid ingredients.

The use of excessive shear when the ointment base (or finished formulation) is cold/thick will cause loss of structure and viscosity. This loss of structure will be seen as oil "bleed" or syneresis as the lighter fractions of the base separate from the bulk formulation on standing.

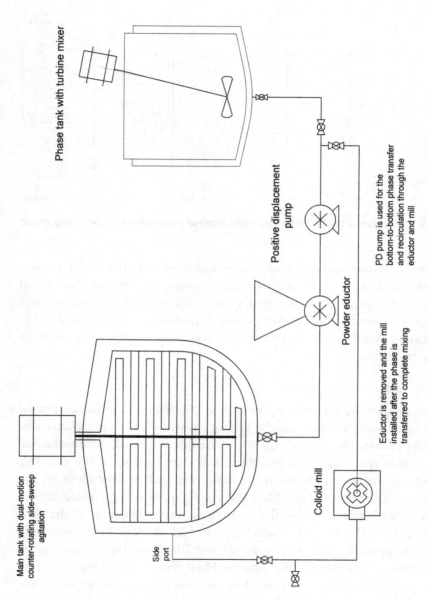

Figure 7.7 Dual motion tank with powder addition, recirculation and external mill, and turbine phase tank.

Phase tank with turbine mixer

Main tank with dual-motion counter-rotating side-sweep agitation

Positive displacement pump

Powder eductor

PD pump is used for the bottom-to-bottom phase transfer and recirculation through the eductor and mill

Eductor is removed and the mill installed after the phase is transferred to complete mixing

Colloid mill

Side port

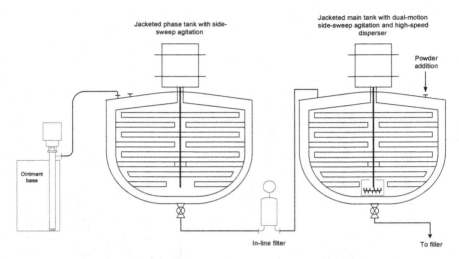

Figure 7.8 Jacketed dual-motion tank with internal disperser and side-sweep phase tank.

Typical ointment bases comprise petrolatum and mineral oil, or petrolatum and waxy/fatty alcohol combinations, the ratio and grades of these components being selected to give the desired finished product viscosity/spreadability. Proprietary ointment bases are also available (e.g., Ointment Base No. 4, 6, and 8 available from Calumet Specialty Products Partners, L.P.).

7.1.5 Gels

Although gels are typically aqueous systems, hydroalcoholic gels can be produced using carbomers as the gelling agent. As is the case when making suspensions, the soluble ingredients are first dispersed and dissolved in the solvent system, and the formulation is thickened up at the end. The mixing geometry employed in the main mix tank is dependent on the viscosity of the finished gel. Thin gels can be made using turbine mixers, possibly with the aid of baffles to increase the turbulence of the flow (Figure 7.5). Thicker gels might require dual-motion counter-rotating blades to provide adequate mixing (Figure 7.6). Solid ingredients can be added via an external powder eductor. High shear can be employed during the addition of the thickener however, once the thickener swells, the level of shear needs to be reduced to prevent loss of structure and viscosity in the final product. Too vigorous mixing during the gelling process can also lead to incorporation of air that can be difficult to remove from the finished gel.

When using Carbomer to produce hydroalcoholic gels, the Carbomer is dispersed in the solvent system prior to forming the gel by addition of a neutralizing agent (for example triethanolamine). Gels, in particular hydroalcoholic gels, can be made from different phases—the aqueous and alcoholic soluble ingredients being dissolved in the component phases prior to bringing them together and forming the finished gel (Figure 7.7).

Aqueous gels usually have a higher viscosity than hydroalcoholic gels, due to the level of hydrogen bonding in the different systems.

7.1.6 Emulsions

Emulsions are thermodynamically unstable two-phase systems consisting of at least two immiscible liquids (e.g., oil and water). The oil and water phases are heated above the emulsification temperature of the system, heat stable ingredients (e.g., preservatives, emulsifier, thickener, and pharmaceutically active ingredients) are dissolved in the respective phases, and the phases are brought together in the main mix tank. At the elevated temperature, the viscosity of both phases is at a practical minimum to make droplet dispersion and emulsion formation easier. The formation of the emulsion requires energy and motion to disperse the two phases. The larger phase, which is originally contained in the main mix tank, usually becomes the continuous phase.

When the oil and water phases are initially brought together, the interface between the two phases is deformed and large droplets are formed. Subsequent mixing breaks up these large droplets. A greater mixing intensity with respect to impeller tip speed creates smaller dispersed phase droplets and a more stable emulsion. The emulsifier (surfactant) helps form and stabilizes the dispersed phase droplets by adsorbing at the interface of the droplets to prevent coalescence. In o/w systems, polymers (e.g., carbomers) thicken and stabilize by forming a gel around the oil droplet. Once the disperse phase droplets are formed and distributed, the emulsion "solidifies" as the temperature is reduced.

After the emulsion "solidifies", further high-shear mixing will cause loss of structure and viscosity in the final product and lead to stability problems. As the emulsion cools, ingredients such as pH adjusters and temperature-sensitive ingredients (e.g., fragrances) are added under low-shear mixing.

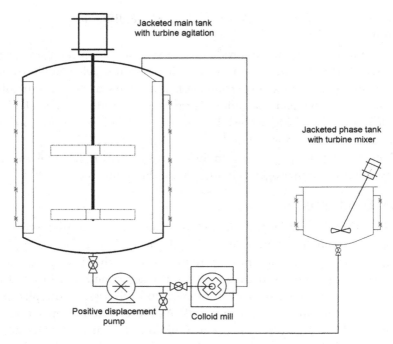

Figure 7.9 Jacketed turbine tank with external mill and recirculation, and jacketed turbine phase tank.

The mixing systems employed in the main mix tank for emulsions depend, to some extent, on the viscosity of the final formulation. Thin lotions might be mixed using a turbine mixer in a baffled tank (Figure 7.9). Thicker creams might need higher torque mixing provided by dual-motion counter-rotating blades or anchor sweep blades (Figures 7.7, 7.10–7.12). The mixing vessels need to be jacketed to allow for controlled heating/ cooling, and have side scrapers to provide efficient heat transfer. Dispersators, homogenizers, and colloid mills are used to help with combining the phases and generating the small dispersed-phase droplets.

As the use of high shear above and low shear below the temperature at which the emulsion "solidifies" is critical to the viscosity and stability of the final emulsion, it is essential that this temperature is determined during initial mixing exercises. This can be accomplished on a small scale by combining the two phases at elevated temperature and slowly cooling the mixture while applying moderate levels of shear. When the emulsion "solidifies", the temperature curve will change (plateau or become variable) when "solidification" occurs.

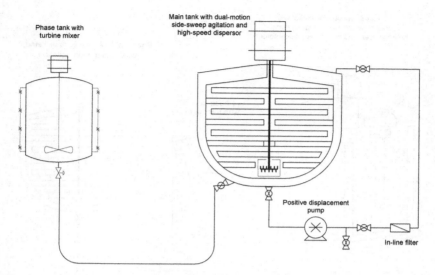

Figure 7.10 Jacketed dual-motion tank with internal homogenizer and recirculation, and turbine phase tank.

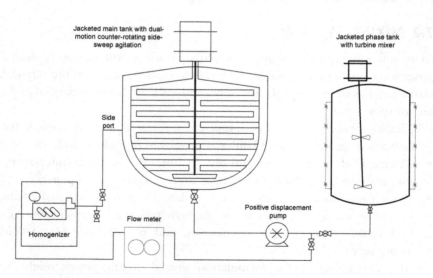

Figure 7.11 Jacketed dual-motion tank with external homogenizer and recirculation, and jacketed turbine phase tank.

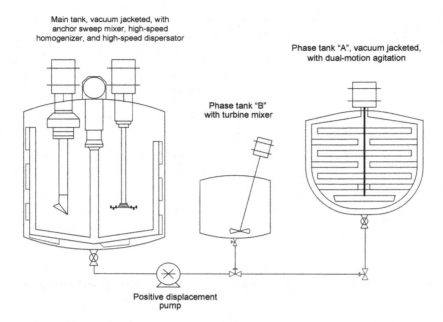

Figure 7.12 Jacketed anchor sweep tank with internal homogenizer and dispersator, phase tanks with dual motion and turbine mixing.

7.2 MIXING SYSTEMS

Many different types of mixing vessels, mixer blade geometries, and external process accessories are available to choose from, depending on the physical characteristics of the formulation in question. Below is a summary of just a few of these options:

- Mixing vessels—open/closed top, pressure rated, heated/cooled, flat bottomed/domed, contain baffles, top/bottom ingredient addition
- Mixing blade geometries—turbine, anchor, side sweep, counter-rotating, saw-tooth dissolvers, rotor-stator homogenizers, wall scrapers
- Process accessories—transfer pumps, in-line filters (e.g., large particulate, polishing, sterilizing), rotor-stator and high-pressure homogenizers, colloid mills, versators, heat exchangers, and powder eductors (avoid agglomeration, "fish eyes")

Also, depending on the formulation, several mixing vessels might be combined within the mixing system. This is often the case when mixing formulations comprised of different phases. The following process diagrams give examples of some typical mixing systems:

- Figure 7.1 shows a temperature jacketed tank with a turbine mixer. This system is used for simple solutions in which the ingredients are added via the top of the tank. Heating and cooling can be used to aid the dissolution of the ingredients and speed up processing. The finished formulation is then transferred from the bottom of the tank through an in-line filter to the filling line or storage vessel with the aid of a positive displacement pump

• Figure 7.2 shows a tank with wall baffles and a turbine mixer. This system is used for simple solutions in which the ingredients are added either via the top of the tank or through the bottom of the tank via a positive displacement pump. The use of wall baffles in the mix tank increases turbulent flow within the tank, thereby improving mixing efficiency. The positive displacement pump is also used to transfer the finished formulation from the bottom of the tank to the filling line or storage vessel

• Figure 7.3 shows a temperature-jacketed main mix tank with side-sweep and turbine agitation. The side-sweep mixing blade continually moves material from the walls of the tank into the bulk, thereby improving heat transfer. The mixing system also has a phase tank with a turbine mixer. The main-tank and phase-tank ingredients are added via the top of the respective tanks. The side phase is transferred from the bottom of the phase tank to the bottom of the main tank via a positive displacement pump. The positive displacement pump is also used to recirculate material from the bottom of the main mix tank, through an in-line filter (to remove particulates above a given size), and back into the top of the tank. This recirculation of material from the bottom to the top of the tank improves mixing. It can also be used to prevent suspended materials settling to the bottom of the tank

• Figure 7.4 shows a temperature-jacketed mix tank with side-sweep and turbine agitation. Ingredients are added via the top of the tank. The finished formulation is transferred via an in-line homogenizer (to reduce particle size) to a storage vessel using a positive displacement pump. The storage tank is equipped with a turbine mixer and wall baffles. Material is recirculated from the bottom of the storage tank, through a heat exchanger, and back into the top of the tank. The heat exchanger is used to remove the heat that has been generated by the homogenization process, and can be used for formulations that have volatile ingredients that need to be kept cool (e.g., fragrances)

• Figure 7.5 shows a temperature-jacketed tank with wall baffles and a turbine mixer. Liquid ingredients can be added via the top of the tank. Powdered ingredients can be added via a powder eductor on an external recirculation loop. Adding powders by this route helps wetting and dispersion, and avoids the formation of "fish eyes"—especially with hygroscopic materials such as thickeners. The powder eductor can also be used to incorporate suspended ingredients into the mix. This process system can be used with simple gels and suspensions

- Figure 7.6 shows a temperature-jacketed tank with dual-motion counter-rotating mixing blades. This process system can be used with gels, suspensions, and ointments. Liquid ingredients (or ointment bases) can be added via the top of the tank. In the case of ointments, heat is used to melt the ointment base. Powdered ingredients can be added via a powder eductor on an external recirculation loop. Heat might also be used to aid the dissolution of powders

- Figure 7.7 shows a temperature-jacketed main mix tank with dual-motion counter-rotating mixing blades. The mixing system also has a temperature-jacketed phase tank with a turbine mixer. The side phase is transferred from the bottom of the phase tank to the bottom of the main tank via a positive displacement pump. Powdered ingredients can be added via a powder eductor during the transfer of the side phase. An external colloid mill is used to reduce the particle/emulsion droplet size to complete mixing. This process system can be used with gels, suspensions, and emulsions

- Figure 7.8 shows an ointment processing system. The main mix tank is temperature jacketed and pressure rated. The mixing system comprises dual-motion counter-rotating mixing blades and an internal homogenizer. A temperature-jacketed, pressure-rated, phase tank with dual-motion counter-rotating mixing blades is also used. Both tanks are equipped with side sweep, to aid heat transfer. The ointment base is pre-warmed and pumped into the phase tank. In the phase tank, the ointment base is further heated and mixed using low shear—to keep the different oil fractions within the base homogenous. When at temperature, the ointment base is pressure fed through an in-line filter to the main mix tank. Once in the main mix tank, the pharmaceutically active ingredients (powders) are added and dispersed with the aid of the homogenizer. Once dispersed, the homogenizer is turned off, and the mixture cooled under low-shear mixing with the dual-motion mixing blades. This system could be used to aseptically produce sterile ointments if the in-line filter was a $0.22\,\mu m$ sterilizing filter, and the powders were presterilized and introduced into the main mix tank in a Class 100 environment

- Figure 7.9 shows a system that would be suitable for producing thin emulsions (lotions, thin creams). A temperature-jacketed main mix tank is used with wall baffles and a turbine mixer. A temperature-jacketed phase tank with a turbine mixer is also used. For making an o/w emulsion, the water would be added to the main mix tank, heated, and the aqueous-phase components added while mixing. The oil-phase ingredients would be added to the phase tank and heated while mixing. When both phases were at the correct temperature and all the ingredients were dissolved, the oil phase would be pumped into the main mix tank using the positive displacement pump. Once transfer was complete and the two phases combined, the solution would be recirculated through the colloid mill to reduce the size of the oil-phase droplets. When the internal phase droplets were the correct size, the mixture would be cooled and milling stopped

- Figure 7.10 shows a system that would be suitable for producing thicker emulsions. In this system, the main mix tank is temperature jacketed, pressure rated, and has dual-motion counter-rotating mixing blades and an internal homogenizer/emulsifier. The phase tank is temperature jacketed with a turbine mixer. With this system, the oil phase is transferred under vacuum to the main mix tank, and enters the tank in close proximity to the homogenizer/emulsifier. Use of this bottom entry port ensures that the oil phase is rapidly dispersed and droplet-size reduction begins during the phase-transfer process. High-torque, low-shear mixing blades with wall scrapers are used in the main mix tank to ensure efficient mixing and heat transfer while not destroying the structure in the emulsion. An external recirculation loop with polishing filter is used to enhance the aesthetic properties of the final emulsion

- Figure 7.11 also shows a system that would be suitable for producing thicker emulsions. In this system, the main mix tank is temperature jacketed and has dual-motion counter-rotating mixing blades. The phase tank is temperature jacketed with a turbine mixer. With this system, the oil phase is transferred to the main mix tank using an in-line pump. Once phase transfer is complete, the formulation is recirculated through an external homogenizer to reduce the oil-phase droplet size. The feed rate through the homogenizer is controlled using an in-line flow meter. As with the previous system, high-torque, low-shear mixing blades with wall scrapers are used in the main mix tank to ensure efficient mixing and heat transfer while not destroying the structure in the emulsion

- Figure 7.12 shows a multiphase emulsion processing system. The main mix tank is temperature jacketed and contains high-torque, low-shear anchor blades with side sweep, an internal high-speed homogenizer and dispersator. The processing system also contains two side phase tanks. One of the phase tanks is temperature jacketed with dual-motion counter-rotating mixing blades, and would be suitable for the oil phase. The other phase tank contains a turbine mixer, and might be used to disperse and suspend powders (e.g., pigments) in a side water phase. The side phases are transferred to the main mix tank using an in-line pump. The high-speed dispersator would be used to quickly and efficiently disperse the suspended phase. The high-speed homogenizer would be used to create the correct internal phase droplet size. Once these processes were complete, the high-torque, low-shear mixing blades with wall scrapers would be used to ensure efficient mixing and heat transfer while not destroying the structure in the emulsion

Notes:

1. The above process system examples are intended to give an idea of the wide range of mixing set-up possibilities. Emulsions, whether o/w or w/o can be produced in a similar fashion. The ability to add external accessories (homogenizers, colloid mills, filters, etc.) increases the flexibility of the tanks.

2. The use of internal homogenizers or recirculation loops with colloid mills/homogenizers, and so on, forms the basis of multipass systems. The formulation passes through the device several times, the reduction in particle or droplet size increases with each pass. The number of times that the formulation will pass through the device (i.e., the number of theoretical passes) can be calculated from the batch size and the pumping rate.

3. If only one pass through a colloid mill or homogenizer is desired, a system whereby the formulation is pumped out of the main mix tank, through the device, and into a storage tank would be used.

7.3 QUALITY BY DESIGN (QbD) FOR FORMULATION AND PROCESS DEVELOPMENT

The goal of formulation and process development is to produce a product in a reliable, consistent, and repeatable way. Finding a stable formulation is only the first step in the development process. Quality by Design is a modern, scientific, and systematic approach to ensure product quality by developing a thorough understanding of the compatibility of the finished formulation to all of the components and processes involved in manufacturing that product. QbD requires identification of all critical formulation attributes and process parameters as well as determination of the extent to which any variation can impact the quality of the finished product.

Having defined the Quality Target Product Profile (QTPP), the critical quality attributes (CQAs) are identified for the product. The QTPP includes the factors that define the desired product and the CQAs include the product characteristics that have the most impact on the product quality. These provide the framework for the product design and understanding.

Beginning with the QTPP, a risk assessment of both formulation ingredients and process steps is undertaken to define the boundary and limitations associated with each material and the process. The combined understanding of the critical material attributes (CMAs) and critical process parameters (CPPs) contributes to meeting the CQAs. Preformulation studies are performed to narrow down formulation choices. A series of experiments are performed to determine ingredient solubility, compatibility, and stability with each formulation type. This comprehensive Design of Experiments (DoE) is conducted to lay the groundwork for minimizing risks at the very beginning of the formulation development process. A statistical analysis of the results of each individual study, using software such as JMP® from SAS Institute, narrows down formulation options based on actual data.

Once formulations have been eliminated from contention, those that remain undergo preliminary "stress" testing. These tests serve to further narrow the options by exposing bench-scale formulations to environmental and other risks. For example, temperature sensitivity tests are completed at both extremes to observe how the formulation reacts. Formulations are heated to +40–50 °C, as well as being taken through a series of freeze–thaw cycles. Another preliminary test would be a centrifugation study to assess how the formulation might react to sitting on a shelf for its projected lifespan.

Having reduced the number of formulation possibilities through stress testing, more extensive informal stability exercises are performed as an initial examination of the remaining formulations to identify and eliminate more obvious stability issues. If a preferred packaging type (e.g., aluminum tube, LDPE bottle) has already been identified in the QTPP, the tests are conducted in that package. If not, the tests are conducted by placing the formulation in a glass container to evaluate its overall stability. Testing encompasses physical (e.g., microscopic appearance, pH, viscosity, specific gravity), chemical (e.g., degradation tests on pharmaceutically active ingredients, preservatives, antioxidants), biological (e.g., activity and performance of pharmaceutically active ingredients), and microbiological (e.g., effectiveness of preservative system) evaluations.

Once a stable formulation has been identified on a bench-scale, the process of scaling up the formulation can commence. During the scale-up stage, feasibility batches are produced to further examine the CMAs and CPPs. This examination measures the extent to which any variation in a component or the process can impact the quality, safety, and efficacy of the finished product. A transition from one scale to another often requires modification of the process—for example, an equipment change may trigger studies to reevaluate the mixing parameters required to reach an optimal viscosity, droplet or particle size. Consequently, the CPPs defined earlier in the development program might need to be reevaluated in light of any change in equipment, material attributes, or process changes. Further DoE studies are conducted during scale up to create the final batch process. Formal, ICH [1] compliant, stability studies (see Section 7.4) are undertaken throughout the scale-up stage to ensure consistent formulations that meet the CQAs.

A schematic summary of the formulation development process is shown in Figure 7.13.

7.3.1 Example of QbD Formulation Development [3]

7.3.1.1 Preformulation

- The QTPP provides information on the intended use, target population, route of administration (oral, topical, vaginal, otic, rectal, ophthalmic, parenteral), dosage form and strength, delivery system and container-closure system
- Preformulation studies characterize the drug substance (pharmaceutically active ingredient) and define the CMAs, evaluate the possible formulation ingredients and methods of handling/mixing. Methods of mixing/handling are compared to existing processes and equipment
- Preformulation studies include
  - solubility studies (see below)—the results of the solubility studies will influence the dosage form and delivery system (e.g., solution or suspension)
  - compatibility experiments between the drug substance and excipients—do any of the proposed formulation ingredients cause a loss in potency of the drug substance
  - initial packaging component and process material compatibility—samples of the packaging components or process material contact parts (e.g., tank wall scraper blades, seals, gaskets, tubing) are immersed in the formulation ingredients (or the proposed formulation). After a given

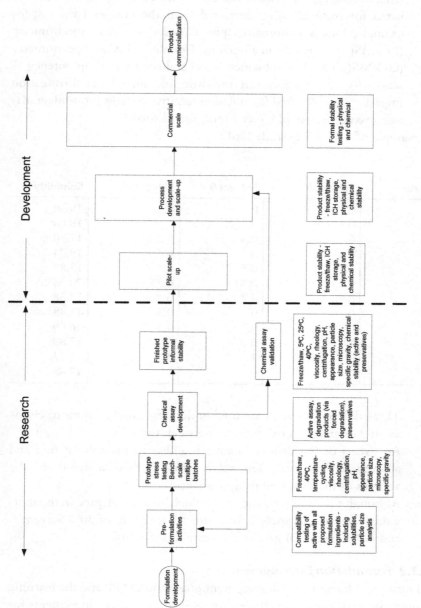

Figure 7.13 Formulation development process.

exposure time, the contact parts are examined for signs of degradation (e.g., swelling, going brittle), and the formulation ingredients are evaluated for material being extracted from the contact parts (by, for example, Liquid Chromatography Coupled to Mass Spectrometry [LC-MS] or Gas Chromatography Coupled to Mass Spectrometry [GC-MS]). The drug substance is also evaluated for loss in potency

- sensitivity investigation—are the drug substance, other formulation ingredients, or finished formulation sensitive to light (photodegradation), oxygen, heat, pH, metal ions, and so forth?

- Example of a solubility study DoE:

Trial run	Solvent A	Solvent B	Solvent C	Solubility
1	0.7	0	0.3	15.30
2	0.5	0.5	0	18.96
3	0	0.3	0.7	17.70
4	0.5	0	0.5	19.24
5	0	0.5	0.5	13.89
6	0.33	0.33	0.33	20.38
7	0.3	0.7	0	23.58
8	0.7	0	0.3	17.48
9	0	0.7	0.3	16.09
10	1	0	0	17.89
11	0	1	0	11.72
12	0	0	1	17.88

- Three different solvents are used in different combinations to evaluate any interactions or synergies between the solvents.
- An excess of the drug substance is added to the solvent mix and allowed to "dissolve". The solvent mix is then analyzed for the amount of the dissolved drug substance.
- Although the drug substance was soluble to some degree in the individual solvents themselves, Trial run #7 (mixture of 30% Solvent A and 70% Solvent B) gave the maximum solubility.

7.3.1.2 Formulation Development

- During the formulation development phase, the QTPP and the learning from the preformulation studies are used to help select ingredients for the finished dosage form. The US Federal Drug Administration (FDA) Inactive Ingredient Guide (IIG) [2] can be referenced for the maximum use levels of inactive ingredients in the desired type of product. Compendial materials are used when available, especially for pharmaceutical and over-the-counter products:

- Emulsifiers/surfactants—for emulsion systems
- Thickeners/stabilizers—for emulsions, gels, and suspensions
- Preservatives—for nonsterile and multiuse products. Multiuse jars of cream might need a more robust preservative than, for example, a pump bottle due to the user repeatedly inserting their fingers into the product to "scoop" it out
- Aesthetic modifiers, depending on the place of application—for example, a face cream cannot be overly greasy
- Buffers/pH adjusters—depending on the sensitivity of the formulation
- Antioxidants—depending on the sensitivity of the formulation
- During formulation development, stress testing is used to give an initial indication of formulation stability (see Section 7.4.1):
  - Phase separation or settling of solid materials during centrifuging—might need to increase the concentration of the emulsifier, use a more robust emulsifier, use a polymeric thickener, or increase the shear while mixing above the emulsification temperature to reduce the internal phase droplet size
  - High-temperature stability issues—might need to decrease the processing temperature or use more temperature tolerant ingredients
  - Separation on freeze–thaw cycling—might need to add a humectant to lower the freezing point of the formulation and thereby reduce the formation of ice crystals
  - Instability of the formulation during extended/exaggerated mixing—waxes and fatty alcohols typically used in emulsions lose viscosity when mixed below their melting point for an extended period of time. This leads to a loss of structure and thinning of the emulsion, and possibly phase separation or settling of solid materials on standing. Using higher processing temperatures, above the emulsification temperature, when high-shear mixing is used to produce the internal phase droplets might overcome this problem

7.3.1.3 Preliminary Risk Assessment
- An initial risk assessment is performed to define the CQAs and assess how they could potentially be affected by the various processing steps.

The CQAs are the product characteristics that have the most impact on the product quality, and need to be controlled to assure a satisfactory formulation is produced each time. Specifications (physical, chemical, biological, and microbiological) are generated with acceptance limits against which each batch of product is tested. Below is an example of an initial risk assessment for the production of a zinc oxide cream. The process steps that are determined as likely having a higher impact on the CQAs are the ones that require more attention during process development and tighter control during subsequent manufacture.

Drug product CQAs	Process steps				
	Water-phase preparation	Oil-phase preparation	Oil-phase addition	Emulsi-fication	Cooling
Description	Low	High	High	High	Medium
pH	Low	Low	Low	Low	Low
Viscosity/Rheology	Low	High	High	High	High
Zinc oxide assay	Low	High	High	Low	High
Preservative assay	Medium	Low	Low	Low	Low

- The preservative assay could potentially be affected if the formulation preservatives are not completely added, dissolved, or uniformly distributed within the final formulation
- All but the pH and preservative may be affected if the oil-phase ingredients are not correct, or if the oil phase is not incorporated uniformly and consistently into the o/w emulsion
- Appearance and physical performance could be compromised if the emulsifiers were incorrect or not prepared properly
- Stability and flow characteristics of the product may be jeopardized if the cooling of the product is too rapid or too slow based on the melting and congealing points of the excipients

7.3.1.4 Stress Testing Assessment

- An assessment of the results of stress testing the above zinc oxide cream was performed. The possible ways to formulate around these instabilities are also listed:
  - Viscosity dropped by 50% post-freeze–thaw cycling—a humectant could be added to lower the freezing point of the formulation, thereby avoiding the formation of large harmful ice crystals
  - Zinc oxide settled out during centrifugation—a polymeric thickener could be added to increase the structure and yield stress of the formulation
  - Large white agglomerations of zinc oxide appeared after extended low-shear mixing—this observation was possibly due to interaction/attraction between the zinc oxide particles. To investigate, a surface-treated grade of zinc oxide could be used to increase repulsion between the particles and help dispersion in the formulation
  - While trying to break up the zinc oxide agglomerations, homogenization caused the product viscosity to drop by 50%

7.3.1.5 Formulation Modifications

- Further development work was undertaken to address the instabilities in the formulation identified during stress testing. Below is an example of a DoE to address these formulation weaknesses:

Trial	Propylene glycol	Acrylate copolymer	Zinc oxide	Mix time
1	+	+	−	−
2	−	−	−	+
3	+	−	+	−
4	−	−	−	−
5	−	+	−	−
6	+	−	+	−
7	+	+	−	+
8	+	−	−	+
9	+	+	+	+
10	−	+	+	+
11	−	−	+	+
12	−	+	+	−

- Propylene glycol: "+" represents formulations with propylene glycol (humectant)
- Acrylate copolymer: "+" represents formulations with acrylate copolymer (thickener)
- Zinc oxide: "+" represents formulations with uncoated USP zinc oxide, "−" represents formulations with surface-treated zinc oxide
- Mix time: "+" represents extended mixing time to disperse the zinc oxide
- A statistical assessment of the formulation modifications with respect to optimizing viscosity and zinc oxide particle size was undertaken (Figure 7.14). In the above example, the addition of acrylate copolymer and propylene glycol is critical to formulation viscosity, and the type of zinc oxide is critical to the particle size in the product
- Based on the results obtained, the CMAs of the drug substance and excipients that most affect the formulation CQAs can be identified. These CMAs in turn define what parameters should be tested/controlled on incoming inspection of the ingredients

7.3.1.6 Informal Stability Testing

- After completing the DoE trials and gaining a solid understanding of the CMAs for the ingredients and the CQAs of the formulation, one or more of the leading formulation candidates are selected for informal stability testing. Physical attributes such as description, pH, viscosity, and specific gravity, and chemical attributes such as drug substance and preservative assays are typically monitored
- Other, more product-specific attributes might also be monitored—such as:
  - Droplet and/or particle size—to assess the criticality of mixing for an emulsion, or the quality of dispersion in a suspension
  - Rheological stress curves—to assess flow of thixotropic formulations
  - Antimicrobial effectiveness testing—batches are made at different preservative levels (e.g., 50, 75, and 100%) to assess the preservative level needed to provide adequate protection and to set the minimum specification limit
  - Photostability

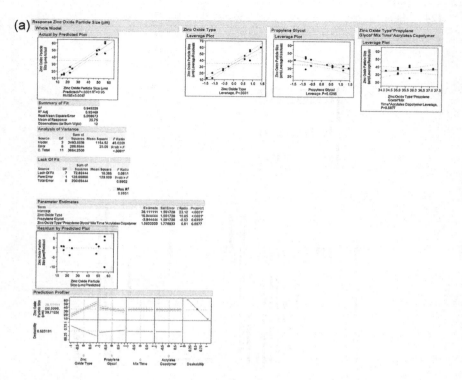

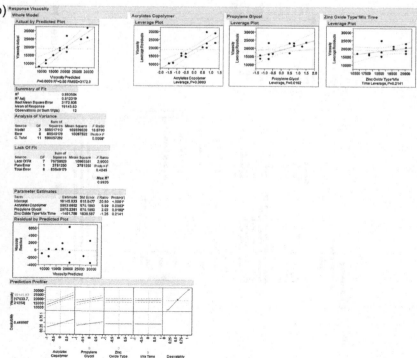

Figure 7.14 (a) Effect of formulation changes on zinc oxide particle size, (b) effect of formulation changes on viscosity, (c) overall statistical assessment of formulation modifications.

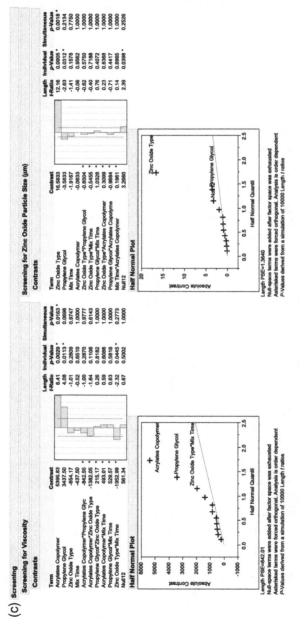

Figure 7.14 Continued

- Below is an example of the results from an informal stability study carried out at 25 °C:

Stability testing example

Test	Time = 0	1 month	2 months	3 months
Description	Pass	Pass	Pass	Pass
pH	5.7	5.6	5.6	5.5
Viscosity	75,000 cps	80,000 cps	82,000 cps	81,000 cps
Specific gravity	1.0	1.0	1.0	1.0
Drug substance assay	102%L	101%L	101%L	100%L
Preservative assay	100%L	100%L	100%L	100%L
Impurities	0.02%	0.03%	0.03%	0.03%

- Informal stability testing is used to help select the optimum formulation(s) to progress to scale-up and, ultimately, commercialization

7.4 STABILITY TESTING

7.4.1 Examples of Physical Instability

- Creaming, flocculation, coalescence, and Ostwald ripening
  This type of physical instability is normally associated with emulsion systems.
  - Creaming occurs when the droplets of the oil phase rise to the surface of the emulsion
  - Flocculation occurs when the droplets of the disperse phase aggregate, but maintain their individual identity
  - Coalescence occurs when the droplets of the disperse phase collide and form larger droplets
  - Ostwald ripening occurs when two droplets of the disperse phase collide to form one larger and one smaller droplet. With continued collisions, the smaller droplets decrease in size until they disappear/become solubilized in the continuous phase. The large droplets continue to grow and eventually separate out (float to the surface)
- Syneresis
  This type of physical instability is seen as liquid separating out from the bulk formula (commonly seen within air bubbles or at the air/product or container/product interfaces).

- Change in viscosity/rheology
  Due to either structure buildup or breakdown within the formulation:
  - Structure buildup (thickening) could indicate crystallization or hardening of waxes, structuring/aligning of polymer chains
  - Structure breakdown (thinning) could indicate micelle coalescence, phase separation, destructuring of polymer chains (effects of salts), polymer chain scission
- Change in particle size
  - Increasing particle size could be indicative of solid particle or emulsion droplet aggregation leading to possible sedimentation or phase separation
  - Decreasing emulsion droplet size could be indicative of Ostwald ripening
- Sedimentation
  Occurs when the thickener/suspending agent used is not "strong" enough to hold particulate matter within the formulation.
- Color change
  Occurs as the visible outcome of certain chemical reactions within the formulation (e.g., reaction of ingredients with each other, oxidation of ingredients, light-induced degradation)
- Change in pH
  Occurs as a measurable outcome of chemical reactions/instabilities between ingredients (or the degradation of ingredients) within the formulation
- Change in conductivity
  A change in the differential conductivity between the top and bottom of a sample with time can be indicative of separation or sedimentation occurring within the sample

7.4.2 Examples of Chemical Instability

- Reduction in drug substance assay/increase in degradation products.
  Due to degradation of the drug substance following chemical reaction with other ingredients in the formulation, or environmental instability (e.g., thermal, oxidative, or light-induced degradation). Some sterilization techniques for sterile products (e.g., autoclaving, gamma irradiation) might also cause degradation of formulation ingredients

• Reduction in the preservative assay.
Due to degradation of the preservative following chemical reaction with other ingredients in the formulation, environmental instability, or being consumed when preserving the formulation

7.4.3 Examples of Biological Instability

• Reduction in biological activity of a drug substance.
Due to degradation or denaturing of the drug substance following the chemical reaction with other ingredients in the formulation, environmental instability, or sterilization incompatibility. The biologically active state of most proteins is characterized by a tightly folded and highly ordered conformation. Chemical or physical denaturants (e.g., urea, acid and heat) can induce the polypeptide chain to unfold.

7.4.4 Examples of Microbiological Instability

• Occurs when the preservative system is "over-powered" by microorganisms—whether introduced during the manufacturing process (from either the ingredients, the personnel making the product, or the equipment) or, more likely, during subsequent use. Bacteria and fungi (yeasts and molds) are the principal types of microorganisms that cause microbiological instability. This instability manifests itself as a change in color, odor (rancidity of fats and oils), curdling, and phase separation in emulsions

7.4.5 Stress Testing

During the initial formulation screening process in which multiple iterations of formulations are produced, stress testing provides a quick and reliable way of comparing the different formulations and screening out physically unstable candidates. Stress testing can also be used during later stages of the development process to give an early assessment of the impact of processing changes (e.g., scale-up, heating/cooling rates, mixing speeds, shear rates, ingredient changes, ingredient concentration changes).

7.4.5.1 Stress Conditions

• Temperature
  • Increasing the storage temperature has the effect of reducing the viscosity of the formulation. This can accelerate the migration of particles and droplets leading to sedimentation or phase separation. Commonly used temperatures are 40 and 50 °C

- Cycling the temperature between ambient room temperature and 40 °C can also exaggerate the above effect
- Freeze–thaw cycling can lead to the formation of ice crystals within aqueous gels and emulsions. This can lead to "cracking" and phase separation. Freeze–thaw testing is conducted by exposing the product to freezing temperatures (approximately −10 °C) for 24 h, then allowing to thaw at room temperature for 24 h (the sample can then optionally be placed in a higher temperature (approximately 45 °C) for 24 h, and then placed at room temperature again for 24 h). The sample is analyzed for significant changes. This completes one cycle. Typically three to five cycles of freeze–thaw testing are evaluated
- Increasing the storage temperature can also accelerate chemical reactions. As a rule of thumb, a 10 °C rise in temperature leads to a doubling in the rate of reaction
- pH
  Measure the pH of the formulation at each time point. Record and graph the pH at each storage condition to identify trends
  Note: pH can only be realistically measured for aqueous systems. pH measurements are temperature dependent
- Conductivity
  Measure the conductivity at the top and bottom of the sample at each time point. Record and graph the difference in conductivity between the top and bottom to identify any trends
- Viscosity
  - Measurements are sensitive to temperature fluctuations, entrapment of air in the sample, channeling, and aggregation of particles or oil droplets in emulsions or suspensions
  - Viscosity measurements can only be compared when using the same instrument, spindle/measuring geometry, and speed
  - Aim for initial readings that are approximately 50% of the measurement range using that spindle/measuring geometry and speed
  - Also, use the next speed setting above and below the one chosen, so that three different readings are obtained. Evaluation of the set of three readings will help identify any sample preparation or experimental error differences as compared to true formulation changes
  - A penetrometer is used for very thick to solid products (measures hardness of sample)

- Rheology (see Chapter 9)
  - Measure the change in rheological properties with time
  - Oscillation stress sweep:
    - characterizes the viscoelastic properties of the material
    - storage modulus (G′) gives a measure of the solid (structural) nature
    - loss modulus (G″) gives a measure of the liquid (viscous) nature
    - Tan δ (=G″/G′)—values of tan δ < 1 are indicative of stability (i.e., product is more solid-like)
    - if G′ decreases with time—indicative of instability
    - if Tan δ increases with time—indicative of instability
    - if the length of the linear viscoelastic region shortens with time—indicative of instability
    - if the cross-over point at which G′ = G″ (i.e., tan δ = 1) reduces with time—indicative of instability
    - these parameters can also be used to compare one formulation or set of processing conditions to another
  - Flow curves
    - used to calculate the yield point (minimum force required to initiate flow)
    - if the yield point decreases with time, the product is flowing more easily and may be indicative of future sedimentation
  - Creep/recovery curves
    - used to determine how deformation occurs over extended time periods
    - can be used to give an indication of long-term sagging or sedimentation effects
- Microscopy
  - Record particle size/oil droplet size, crystal shape/morphology, presence of crystals/particulate matter, and monitor any changes with time
  - A hot stage can be used to evaluate the melting points of any particulate matter. Compare to melting points of ingredients
- Particle size analysis
  - Gives the % population of particles that are within specified size ranges
  - Data also gives mean particle size, and the graphical data can be used to determine whether the particles fall within a single or multiple distributions

- Monitoring data over time gives information relating to crystal growth within the formulation
- This type of data can also be used to compare differences in processing (e.g., oil droplet or crystal size vs. different mixer blades)
- Specific gravity (density)
  - Pycnometer: used for non-Newtonian fluids (semisolids, creams, lotions, etc.)
  - Densitometer: used for Newtonian fluids (liquids)
  - Can be used to assess effects caused by different manufacturing processes (e.g., entrainment of air may give rise to oxidative instability)
- Photostability
  - Two light sources for photostability testing are specified in the ICH Q1B guidance for photostability testing of new drug substances and products [1].
  - This guidance document for photostability testing establishes requirements for exposure times. Confirmatory studies require that samples be exposed to light providing an overall illumination of 1.2 million lux hours and integrated near-ultraviolet energy of not less than 200W h per square meter.
  - Photostability testing is typically conducted at 25 °C
  - Used to accelerate possible chemical instabilities/interactions
- Centrifugation
  - Exposes the sample to extreme force
  - Accelerates possible phase separation and sedimentation effects
  - Typical centrifugation settings are 3500 rpm for 30 min
- Wick test
  - Small sample applied to a piece of filter paper
  - Used to evaluate syneresis
- Mixing sweep test
  - Exposes sample to an extended period of slow, low-shear mixing
  - Used to investigate coagulation of emulsion droplets
- Internal pressure of aerosols
  - Use a pressure gauge with a needle adaptor
  - Measures internal can pressure
  - A decrease in internal pressure could indicate formulation or packaging problems
- The table below gives a guide as to what stress tests might be applicable to the different types of formulation:

	o/w emulsion (creams, lotions)	w/o emulsions (creams, lotions, pastes)	Ointments	Aqueous gels	Anhydrous gels	Liquids (solutions, suspensions)	Aerosols
Temperature	X	X	X	X	X	X	X
pH	X	X		X		X	
Conductivity	X	X		X			
Viscosity	X	X	X	X	X	X	X
Rheology	X	X	X	X	X	Suspensions—creep and yield	
Microscopy	X	X	X	X	X	X	X
Particle size analysis	X	X	X	X	X	X	X
Specific gravity (density)	X	X	X	X	X	X	
Photostability	X	X	X	X	X	X	
Centrifugation	X	X	X	X	X	X	
Wick test	X	X	X	X	X		
Mixing sweep test	X	X	X	X	X		
Internal can pressure							X

7.4.6 Informal Stability Testing

Informal stability testing is used to help select the optimum formulation(s) to progress to scale-up and, ultimately, commercialization. The leading candidates are set up on stability at a range of temperatures (for example 5, 25, and 40°C), and the chemical and physical attributes monitored over time. Typically, a three- to six-month storage period is used, the product being evaluated on a monthly basis. Different temperatures are used above and below the intended storage temperature for the product—the lower temperature acting as a control for chemicals stability, and the higher temperature being used to accelerate any chemical reactions.

The testing performed as part of the informal stability would typically include elements from the stress testing listed in Section 7.4.5.1, as well as analytically determining the level of drug substance and preservatives. Analytical assays need to have been developed by this stage of the project to enable this evaluation.

During or in parallel to informal stability testing, antimicrobial effectiveness testing of the candidate formulations is evaluated. Batches made at different preservative levels (e.g., 50, 75, and 100%) are inoculated with a variety of standardized microorganisms. Following incubation, the effectiveness of the preservative against the microorganisms is determined. USP<51> gives details of the antimicrobial effectiveness test, the microorganisms to be used and their standardization, incubation times and temperatures, and the performance requirements for the preservative system to be deemed acceptable. Making batches at a range of different preservative concentrations enables the minimum level needed to provide adequate protection to be determined. This value can also be used to set the minimum specification limit.

7.4.7 Formal Stability Testing—ICH Guidelines

Formal stability testing is used to provide data (a) to support the shelf-life of the product, (b) to support regulatory submissions, and (c) to support short-term excursions outside the label storage condition. The ICH stability testing guidelines [1] provide details on storage temperatures and testing frequency for different product types, different packaging types, and intended markets (different climatic zones). Testing covers physical and chemical attributes, as well as microbiological and biological evaluation—as applicable to the product. By the time formal stability is set up, the product specifications should be established, and the analytical assays should be finalized and validated.

7.4.7.1 Testing Frequency (ICH Q1A(R2) [1])

For products with a proposed shelf-life of at least 12 months, testing at the long-term storage condition should normally be every 3 months over the first year, every 6 months over the second year, and annually thereafter through the proposed shelf-life.

At the accelerated storage condition, a minimum of three time points (including the initial and final time points) from a 6-month study is recommended.

A minimum of four time points (including the initial and final time points) from a 12-month study is recommended for the intermediate storage condition. Testing at the intermediate storage condition is required if a "significant change" is recorded at the accelerated storage condition. In general, a "significant change" for a liquid or semisolid dosage form is defined as:

- A 5% change in assay from its initial value or failure to meet the acceptance criteria for potency when using biological or immunological procedures
- Any degradation product exceeding its acceptance criterion
- Failure to meet the acceptance criteria for appearance, physical attributes, and functionality test (e.g., color, pH, phase separation, resuspendibility, dose delivery per actuation); however, some changes in physical attributes (e.g., softening of suppositories, melting of creams) may be expected under accelerated conditions

7.4.7.2 Storage Conditions (ICH Q1A(R2) [1])

In general, a drug product should be evaluated under storage conditions that test its thermal stability and, if applicable, its sensitivity to moisture or potential solvent loss.

- General case

Study	Storage condition
Long term	25 °C ± 2 °C/60% RH ± 5% RH, or
	30 °C ± 2 °C/65% RH ± 5% RH
Intermediate[a]	30 °C ± 2 °C/65% RH ± 5% RH
Accelerated	40 °C ± 2 °C/75% RH ± 5% RH

[a]If 30 °C ± 2 °C/65% RH ± 5% RH is the long-term condition, there is no intermediate condition.

- Drug products packaged in semipermeable containers. Aqueous-based products packaged in semipermeable containers should be evaluated for potential water loss in addition to physical, chemical, biological, and microbiological stability. This evaluation is normally carried out under conditions of low relative humidity. A 5% loss in water from its initial value is considered a significant change for a product packaged in a semipermeable container after an equivalent of 3 months' storage at $40\,°C \pm 2\,°C$/not more than 25% RH.

Study	Storage condition
Long term	$25\,°C \pm 2\,°C$/40% RH ± 5% RH, or
	$30\,°C \pm 2\,°C$/35% RH ± 5% RH
Intermediate[a]	$30\,°C \pm 2\,°C$/65% RH ± 5% RH
Accelerated	$40\,°C \pm 2\,°C$/not more than 25% RH

[a]If $30\,°C \pm 2\,°C$/35% RH ± 5% RH is the long-term condition, there is no intermediate condition.

- Drug products intended for storage in a refrigerator

Study	Storage condition
Long term	$5\,°C \pm 3\,°C$
Accelerated	$25\,°C \pm 2\,°C$/60% RH ± 5% RH

If the drug product is packaged in a semipermeable container, appropriate studies should be undertaken to assess the extent of water loss—for example, weight-loss studies.

- Drug products intended for storage in a freezer

Study	Storage condition
Long term	$-20\,°C \pm 5\,°C$

- Drug products intended for storage below $-20\,°C$. Drug products intended for storage below $-20\,°C$ are treated on a case-by-case basis.

7.4.7.3 Short-term Excursions Outside the Label Storage Condition

Examples of when short-term excursions outside the label storage condition may be encountered include during the transport of the product from the manufacturer to the retailer/distributor (i.e., somewhere within the product supply chain), or on the retailer's/distributor's shelf (in the case of adverse storage events—e.g., power outages).

An example of a study to evaluate short-term excursions from the labeled storage conditions is presented below:

- Stress conditions: Two sets of samples are subjected to the following thermal stress conditions:
  - Temperature cycling from 25 to 50 °C (hold at 50 °C for 4 h), to 4 °C (hold at 4 °C for 4 h), then back to 25 °C
  - Store at −20 °C for 2 days, followed by 40 °C for 2 days
  - Store at low pressure (~250 mmHg) for 4 h (25 °C)
  - Store at low pressure (~250 mmHg) for 4 h (25 °C), then cycle temperature (50 °C for 4 h, 4 °C for 4 h)
- Thermal cycle testing: One set of samples is then tested for physical and chemical attributes, for example:
  - Appearance
  - pH
  - Viscosity
  - Particulate matter (presence of crystallization)
  - Drug and preservative assay (including formation of degradation products)
- Packaging robustness testing: The second set of samples is then subjected to vibration and drop testing to assess any adverse effects on the packaging (see Section 11.1.18)
  - Vibration testing (ASTM D 999-08)
  - Drop testing (ASTM D 5276-98)

REFERENCES

[1] ICH Quality Guidelines. Stability Q1A–Q1F, www.ich.org.
[2] US Food and Drug Administration Inactive Ingredients database, www.fda.gov.
[3] Lowenborg M. (R&D Formulation Development Manager, DPT Laboratories Ltd.). QbD based formulation development – the DPT Labs approach. Website presentation, www.dptlabs.com.

CHAPTER 8

Particle Size Analysis: An Overview of Commonly Applied Methods for Drug Materials and Products

Contents

8.1 INTRODUCTION

An understanding of the particle size of semisolids, suspended liquids, or dry powder formulations is one of the critical control parameters used in product development and stability studies for pharmaceutical products. Particle size and morphology of the drug substance or finished product are critical quality attributes that can influence manufacturing processes, stability of the drug product, dissolution rate, drug release rate, dose content uniformity, rheological properties, and grittiness of drug products [1,2]. Particle-size range varies considerably between different types of formulation and routes of administration. Generally, particles for liquid dosages, injections or parenterals are submicron and in the range 1–500 μm for oral suspensions. Semisolids (gels, creams, ointments, lotions), administered topically or transdermally, range from submicron to 200 μm. Orally inhaled and nasal drug products (OINDPs), including dry powders and liquid nasal sprays, range from 1 to 300 μm.

Often particle size is not expressed as just one number but as the size-distribution profile—typically expressed at three threshold levels, for example, d10, d50, and d90 corresponding to the undersize diameter at the tenth, fiftieth, and ninetieth percentiles, respectively.

Essential Chemistry for Formulators of Semisolid and Liquid Dosages
http://dx.doi.org/10.1016/B978-0-12-801024-2.00008-X
137

In the pharmaceutical field, particle-sizing techniques broadly fall into three categories: quantitative separation by inertial methods (cascade impactor, analytical sieving); diffraction of laser light; and imaging (optical microscopy and scanning electron microscopy) [3–5]. The most commonly used techniques are laser diffraction (static and dynamic laser-light scattering) and optical microscopy for both size distribution and morphology of particles. The cascade impaction technique is used for determining the aerodynamic particle size for OINDPs.

8.2 LASER DIFFRACTION

Laser diffraction is the most widely used technique for determining the particle-size distribution of drug materials and drug products. Put simply, incident laser light is scattered by particles, and the scattered light is detected by a series of concentric detector rings. The scattering angle and intensity of the scattered light depend on the size (volume) of the particle—smaller particles scatter light at larger angles with weak intensity, and larger particles scatter light at low angles with stronger intensity. In commercial instruments, the scattered (diffracted) light is used to calculate the particle size based on the concept of "spheres of equivalent volume." Modern laser diffraction instruments can measure the size of particles ranging from 0.01 to 3500 µm. In static light scattering, the signal received by the detector is static or stable. For particles <4 µm, however, dynamic laser-light scattering (DLS) is more suitable. In DLS, the particle size is determined by the Brownian motion of the particles. As small particles move faster than larger particles, the Brownian motion causes the intensity of the diffracted light to fluctuate. The instrument measures these fluctuations in scattering intensity and associates it with particle size. This technique is also known as photon correlation spectroscopy or quasi-elastic light scattering.

8.2.1 Development and Validation of Laser Diffraction Particle-Size Methods

Development of the test method is generally followed by validation. USP <1225> ("Validation of Compendial Procedures") provides guidance on validating test methods, typically, a particle-size test falls into Category III. The most critical parameter when validating a particle-size test method is the precision. USP <429> "Light Diffraction Measurement of Particle Size" provides guidance on precision requirements for particle-size measurement. Generally, for pharmaceutical materials and products, particle size is expressed at three threshold levels—D10, D50, and D90 (also represented as d0.1, d0.5, and d0.9, respectively). These threshold levels indicate undersize particles for the tenth, fiftieth, and ninetieth percentiles, respectively (i.e., the percent of particles finer than the reported diameter; e.g., D10 = 5 μm indicates 10% of the particles are finer than 5 μm). The size distribution data determined by laser diffraction are typically reported as a volume-based analysis—that is, the data represent the diameter of a sphere the volume for which is equivalent to the volume of the particle. The size distribution is determined from the measured value of "Volume Weighted Mean D[4,3]" (also known as the De Brouckere Mean Diameter). D[4,3] represents the mean calculated by $D = \Sigma(d_1{}^4 + \dots d_n{}^4)/\Sigma(d_1{}^3 + \dots d_n{}^3)$, in which d = the diameter of the individual particles. The D[4,3] provides the mean diameter assuming that all of the particles measured fall in a single mode. Therefore, if particle size is to be represented by a single number for a product, the D[4,3] will be the value. However, the size distribution data of the particles (D10, D50, and D90) is more informative and provides useful information relating to the percentage of fine and coarse particles.

Laser diffraction can be used for testing dry as well as wet particles. Testing dry powders is relatively quick and does not involve sample preparation—the sample is simply dispersed in air. However, certain powders (e.g., those that are sensitive to moisture or are hygroscopic) may not be suitable for testing by general dry powder methods. Handling of dusty material is hazardous, and extra caution is warranted. Materials that are static also cause concerns as they are likely to stick within the instrument (e.g., to the tube walls and cell windows) causing poor reproducibility of the data. If the powder does not readily form a uniform dispersion in air, force needs to be applied to aid dispersion. With some crystalline materials, this additional force could break the powder into smaller particles due to the applied vibration, the feeding mechanism, air pressure, and feed velocity. Wet dispersion methods, on the other hand, require a degree of sample preparation (sometimes rigorous sample preparation) and, in some cases, may require the use of a specific dispersant.

8.2.2 Method Development and Validation

Method development and validation activities typically involve:
* Verification of instrument performance: Although laser diffraction is an absolute method requiring no "calibration," the instrument itself employs various mechanical actions and components—including sample circulation, laser alignment, electronics, detectors, and software. Therefore, as part of method development and validation, it is critical to verify the normal

performance of the instrument as a single entity. This is accomplished using standard reference materials—typically NIST (National Institute of Standards and Technology) traceable standard particles are used. USP <429> indicates that if the measured median particle size (D50) for a standard particle falls within ±3% of the certified range, the performance of the instrument is considered acceptable (below 10μm requirement is doubled).

• Selection of the dispersant media: The media selected must be neutral to the test material—meaning the test material should not react, dissolve, or swell in the media. Water is the most common medium; however, for materials that dissolve or swell in water, a different medium is required.

• Sample preparation and stability of the sample: Wet sample methods require preparation of the sample, such that it can be loaded into the dispersant tank for testing. When raw materials (e.g., the drug substance) are suspended in aqueous media, the suspension may require rigorous mixing and brief sonication to homogeneously suspend the individual particles. The suspension media may contain small amounts of surfactant for wetting the material, and to keep it suspended. Samples of solid suspended particles tend to settle quickly; therefore, it is critical to test the stability of the sample and specify the time frame within which samples have to be loaded into the dispersion tank for testing.

• Robustness: Robustness testing establishes that the measured particle size is independent of minor changes to the method parameters. One way of demonstrating robustness is to take measurements at different obscuration levels (the obscuration level is related to the amount of sample added to the dispersion media and, therefore, the final concentration of the material in the dispersant). High obscuration levels can result in backscattering of the light and give an artificially reduced particle size. By testing the particle size at different obscuration levels, an optimum obscuration level can be determined for the test material.

• Intermediate precision: Intermediate precision testing establishes that the data generated by different analysts on separate days are equivalent. Size distribution data from two separate analysts generated on separate days on the same lot of material are compared. The acceptance criteria listed in USP <429> for precision are that the RSD (Relative Standard Deviation) at D50 is ≤10%, and at D10, and D90 ≤15% for six separate measurements. This RSD requirement is doubled at all thresholds when D50 is smaller than 10 μm. To assess intermediate precision, the USP <429> precision criteria can be applied for data from individual analysts, as well as their pooled data.

- Ruggedness: Ruggedness testing shows that results generated by the developed method are applicable to different instruments of the same make and model, possibly located in two different laboratories and operated by different analysts. This is helpful when the method is transferred from one laboratory (the "method developing laboratory") to another (the "method receiving laboratory"). The USP <429> precision criteria may be applied to establish ruggedness of the method. Before testing the developed method for a given material or product, system qualification may be performed by testing any reference material (e.g., standard material) on the two instruments in question. For the reference material, the D50 value obtained at the "method receiving laboratory" should not deviate by more than 10% from the D50 data from the reference laboratory ("method developing laboratory"), and the variations at D10 and D90 thresholds should be no more than 15%.

8.2.3 Concerns about Laser Diffraction Technique

- One of the major concerns for particle-size measurements by laser diffraction is that the data give the diameter of a sphere the volume for which is equivalent to the volume of the particle tested. Whereas this approach provides accurate data when the test particles are spheres or spheroidal, in the real world most drug materials and excipients are nonspherical and have a high aspect ratio (i.e., the ratio of the long axis to the short axis of the particle). For nonspherical particles, the "diameter of sphere of equivalent volume" may not be an accurate measure of the real particle size. This may lead to the generation of data of high precision, but may not be accurate for some materials.
- Another critical factor is that the particle-size distribution data obtained for the same lot of test material (not the reference standards) using instruments from different manufacturers will probably not be comparable. Thus, the size distribution data appear to be dependent on the manufacturer of the instrument, suggesting a need for harmonization between the instrument manufacturers. In the real world, therefore, it becomes imperative that the same instrument make and model is used throughout the life of the product when testing for compliance with established specifications, or when methods are transferred to another laboratory.
- When using the Mie theory for determining particle size by light scattering, accurate information on the refractive index and absorbance index of the test material is needed. This information may not be available for some pharmaceutical actives or excipients.

- When testing products such as creams, emulsions, or lotions, the sample might need to be diluted. Diluting the sample might change the stability of the product itself, thereby limiting the suitability of the light-scattering technique for certain creams and emulsion systems.
- The light-scattering technique is not material specific, and the size data generated cover all the particles or particle-like matter present in the sample—including air bubbles or foam. If the product has two or more types of solid particles (e.g., a mixture of drug material and excipient, both in solid form), the technique cannot be used by itself to obtain size-distribution data specific to an individual type of material present in the sample. For these types of samples, particle sizing coupled to a method of particle identification (e.g., Raman spectroscopy) can be employed. This problem can also be encountered with semisolid products such as topical gels or oral suspensions. Here, the laser diffraction technique on its own may not be suitable as it cannot distinguish between particles of drug substance and the gel particles.

8.3 PARTICLE SIZE EVALUATION BY MICROSCOPY

The use of optical microscopy for determining particle size is common, and had been used for many years before the laser diffraction instruments came into practice. The microscopy technique provides a size distribution based on numerical average, and is also a very useful technique to understand the morphology of the particles. It has been established that for dry powder inhalation formulations, the particle morphology (aspect ratio, roundness, etc.) influences the performance of the product [6–8]. As a result, knowing both the particle-size distribution and particle morphology is critical for certain dosage forms.

Determining particle size by microscopy, however, has its own challenges. The most important factor is preparation of the sample slide. If particles are well separated from each other the software can be used to determine particle count, particle size (as ferret mean, mean diameter, caliper length and width, inscribed and circumscribed circle diameters, etc.), and morphological information (including roundness, aspect ratio, shape, and surface texture).

Calibration verification for microscopes can be achieved using standard particles. USP <776> provides a basic description of the shapes and different diameters for particles encountered in pharmaceutical applications. Optical microscopy is a useful tool for determining the size of particles suspended in gels, creams, or ointments. If two types of solid materials are suspended in the semisolid formulation, it may not be possible to distinguish between the different particles unless the two materials have distinct physical features. In cases in which the two suspended materials in a formula have significantly different melting points, microscopy using a hot stage may be used to distinguish between the two types of materials by their melt temperatures.

Published reports detail ingredient-specific particle sizing using chemical imaging microscopy [9,10]. Techniques using Raman spectroscopy for identifying specific ingredients and then determining the size distribution of the two (or more) types of particles is very impressive. The applicability of this technique is not yet widespread, and mostly it is useful for OINDPs. If the different types of suspended particles are embedded in a matrix (e.g., in gels or ointments), the signal from the particles of interest is generally poor, and data may not be conclusive. A technique using dynamic image capture (employing a high-speed camera of 500 frames per second and an exposure time of 1 ns) and analysis has been reported for the measurement of the particle-size of pharmaceutical excipients [11]. Yu and Hancock have reported that a good correlation was observed between the size distribution of spherical particles determined by the laser diffraction and dynamic image analysis methods, however significant differences arose for rod-like particles [11].

One factor of paramount importance for particle-size determination relates to the way that the test material or the product is sampled [12,13]. It is important to have a sampling plan that will provide a true representation of the bulk material. "Sampling thieves" are used for sampling at different levels/depths from containers of a bulk material. "Spinning rifflers" are used for channeling flowing powder into different sampling vials for testing. A statistical determination of the number of samples needed to achieve the intended precision is essential to provide an accurate determination of particle size.

8.4 CONCLUSION

In general, using just one method of particle characterization may not be sufficient to understand product performance and stability. Therefore when developing pharmaceutical products, both laser diffraction and microscopy techniques may be useful to provide complementary data to give a better understanding of the product.

REFERENCES

[1] Shekunow BY, Chattopadhyay P, Tong HHY, Chow AHL. Particle size analysis in pharmaceutics: principles, methods and applications. Pharm Res 2007;24:203–27.

[2] Fincher JH. Particle size of drugs and its relationship to absorption and activity. J Pharm Sci 1968;57(11):1825–35.

[3] Burgess DJ, Duffy E, Etzler F, Hickey AJ. Particle size analysis: AAPS workshop report, cosponsored by the Food and Drug Administration and the United States Pharmacopeia. AAPS J 2004;6(3). Article 20:1–12.

[4] Brittain HG. Particle size distribution, part-III: determination by analytical sieving. Pharm Technol December 2002;26(12):56–64.

[5] Iacocca R. Particle size analysis in pharmaceutical industry. Am Pharm Rev 2007; 10(May/June):18–23.

[6] Kaialy W, Alhalaweh A, Velaga SP, Nokhodchi A. Effect of carrier particle shape on dry powder inhaler performance. Int J Pharm 2011;421(1):12–23.

[7] Mack P, Horvath K, Garcia A, Tully J, Maynor B. Particle engineering for inhalation formulation and delivery of biotherapeutics. Inhal Mag August 2012.

[8] Kaialy W, Alhalaweh A, Velaga SP, Nokhodchi A. Influence of lactose carrier particle size on the aerosol performance of budesonide from a dry powder inhaler. Powder Technol 2012;227:74–85.

[9] Doub WH, Adams WP, Spencer JA, Buhse LF, Nelson MP, Treado PJ. Raman chemical imaging for ingredient-specific particle size characterization of aqueous suspension nasal spray formulations: a progress report. Pharm Res 2007;24(5):934–45.

[10] Kuriyama A, Ozaki Y. Assessment of active pharmaceutical ingredient particle size in tablets by Raman chemical imaging validated using polystyrene microsphere size standards. AAPS Pharm Sci Tech 2014;15(2):375–87.

[11] Yu W, Hancock BC. Evaluation of dynamic image analysis for characterizing pharmaceutical excipient particles. Int J Pharm 2008;361:150–7.

[12] Brittain HG. Particle size distribution, part-II: the problem of sampling powdered solids. Pharm Technol July 2002;26(7):67–73.

[13] WHO guidelines for sampling of pharmaceutical products and related materials. In: World Health Organization Tech. Report. 2005. Sr. No. 929.

CHAPTER 9

Rheological Studies

Contents

Rheology is the study of how materials deform and flow as a result of an external force, and is one tool that is used by formulators to help characterize semisolid dosage forms [1]. The two extremes of rheological behavior are:

- Elastic behavior—in which any deformation reverses spontaneously when the applied force is removed. Energy is stored by the system, then released (e.g., for rigid solids). For an elastic solid, stress is proportional to strain;
- Viscous (or plastic) behavior—in which any deformation ceases when the applied force is removed. Energy performs work on the material and is dissipated through the system (e.g., for Newtonian liquids). For a viscous fluid, stress is proportional to strain rate.

Essential Chemistry for Formulators of Semisolid and Liquid Dosages
http://dx.doi.org/10.1016/B978-0-12-801024-2.00009-1

Between elastic and viscous behavior lies the real world of most substances, which are viscoelastic materials. In the case of viscoelastic materials, timescales are important to rheological behavior—when deformation is fast, materials exhibit solid-like behavior; when deformation is slow, materials exhibit fluid-like behavior.

The rheology of semisolid formulations provides critical information for product and process performance, including product stability. Rheology-based measurements can help predict which formulations might exhibit flocculation, coagulation, or coalescence, resulting in undesired effects, such as settling, creaming, or separation. Viscosity is a measure of the resistance of a material to flow, and is just one aspect of the rheological profile of a formulation [2].

This chapter looks at rheological terms that are commonly used with semisolid formulations, examples of instruments, measuring geometries, and different experimental procedures.

9.1 GENERAL TERMS AND DEFINITIONS [3]

- Shear flow—deformation as a result of an external force
- Shear rate ($\dot{\gamma}$)—a measure of the speed of a shear flow resulting from the application of a shear stress to a fluid (s^{-1})
- Shear stress (τ)—the stress (force per unit area) that causes successive parallel layers of a material to move relative to each other (Pa)
- Shear strain (γ)—a normalized measure of deformation (= [change in length]/[original length]). When deformation occurs, different parts of a body are displaced by different amounts
- Flow curve—a graphical representation of the behavior of flowing materials in which shear stress is related to shear rate
- Viscosity (η)—a measure of the resistance to flow. In shear deformation, viscosity is the ratio of applied shear stress to resulting shear rate ($\eta = \tau/\dot{\gamma}$). Viscosity is typically reported in units of Poise (P), centiPoise (cP), Pascal seconds (Pa s), or milliPascal seconds (mPa s). Viscosity is a single-point measurement on the rheological flow curve
- Viscoelasticity—the phenomenon of exhibiting both elastic (solid-like or energy storing) and viscous (liquid-like or energy dissipating) properties. A viscoelastic material will exhibit viscous flow under constant stress, but a portion of mechanical energy is conserved and recovered after the stress is released
- Newtonian fluid (Figures 9.1 and 9.2)—a fluid with a viscosity that is independent of the shear conditions

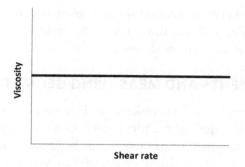

Figure 9.1 Newtonian fluid.

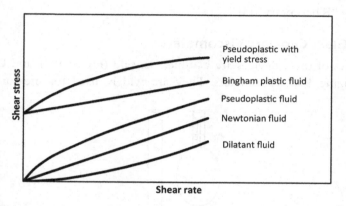

Figure 9.2 Example flow curves for Newtonian and non-Newtonian fluids.

- Non–Newtonian fluid (Figure 9.2)—a fluid that exhibits viscosity that is dependent upon the shear conditions. Non–Newtonian fluids may also exhibit yield stress. For example:
    - Shear thickening (dilatant)—viscosity increases with the rate of shear stress
    - Shear thinning (pseudoplastic)—viscosity decreases with the rate of shear stress
    - Thixotropic—becomes less viscous over time when shaken, agitated, or otherwise stressed
    - Rheopectic—becomes more viscous over time when shaken, agitated, or otherwise stressed
    - Bingham plastics—behave as a solid at low stresses but flow as a viscous fluid at high stresses

- Yield stress—a critical shear stress value below which a material behaves like a solid (i.e., will not flow). Once the yield stress is exceeded, the material yields (deforms) and flows.

9.2 INSTRUMENTS AND MEASURING GEOMETRIES

The most commonly used viscometers and rheometers for semisolid formulations are glass capillary and rotational viscometers, and rotational rheometers. The rotational instruments can be stress controlled or rate controlled. A variety of measuring geometries are available for the mechanical instruments depending on the physical nature of the material being tested. The use of these instruments is summarized in USP General Chapter <1911> "Rheometry" [4].

9.2.1 Glass Capillary Viscometers

Examples of these devices are U-tube, Ostwald (Figure 9.3), and Ubbelohde (Figure 9.4) viscometers. They are widely used for measuring the

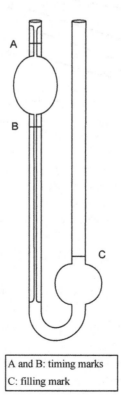

A and B: timing marks
C: filling mark

Figure 9.3 Ostwald glass capillary viscometer.

viscosity of Newtonian fluids, including dilute solutions and suspensions, and are the simplest and least expensive viscometric systems commercially available. They consist of a U-shaped glass tube held vertically in a controlled temperature bath. The time required for a given volume of fluid to flow through a defined length of glass capillary under its own hydrostatic head is measured. By multiplying the time taken by the conversion factor of the viscometer (provided with the unit), the kinematic viscosity is obtained.

9.2.2 Rotational Instruments [3]

In rotational methods, the test fluid is continuously sheared between two surfaces, one or both of which are rotating. These devices are able to shear the sample for an unlimited period of time, so transient or equilibrium-state behavior can be monitored under controlled conditions. Rotational methods can be used to measure the viscosity of fluids, or to characterize the viscoelastic properties (using oscillatory methods).

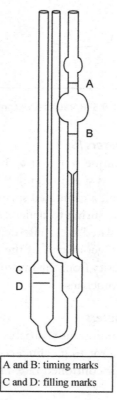

A and B: timing marks
C and D: filling marks

Figure 9.4 Ubbelohde glass capillary viscometer.

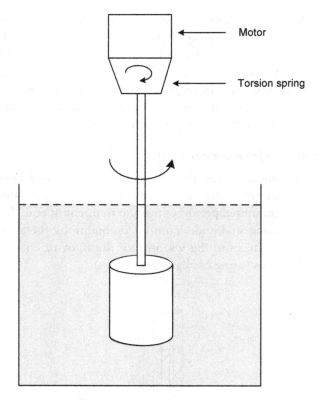

Figure 9.5 Rotational viscometer.

9.2.2.1 Rotational Viscometers [5]

This device measures the torque required to rotate an immersed element ("spindle") in a fluid at a fixed speed (Figure 9.5). The spindle is driven by a synchronous motor through a calibrated spring. Viscous drag of the fluid against the spindle causes the spring to deflect, and this deflection is correlated with torque. The calculated shear rate depends on the speed of rotation and the size and shape (geometry) of the spindle. Conversion factors are needed to calculate viscosity from the measured torque, and are typically precalibrated for specific geometries.

9.2.2.2 Rotational Rheometers

These are high-precision, continuously variable shear instruments in which the test fluid is sheared using rotating cylinders, cones, or plates, under conditions of controlled stress or controlled rate. The rotational system consists of four parts: (1) the measuring geometry; (2) a device to apply a constant torque, oscillatory stress, or rotation speed to the measuring geometry over a wide range of shear stresses or rates; (3) a device to determine the response of the test fluid; and (4) a means of controlling the temperature of the test fluid and measuring geometry (e.g., water bath or Peltier element).

Controlled stress rheometers provide several measurement modes for evaluating materials (these will be discussed in more detail in Section 9.4) [6]. These include:

- Flow—in which the shear stress on the material is increased and then decreased in a controlled manner and the material response is measured. This generates a flow curve showing how the material behaves under conditions of increasing stress, and then how the material "recovers" as the applied stress is backed off
- Creep—in which a constant stress is applied and the material response (displacement) over medium to long time periods is measured
- Oscillation—in which a small sinusoidal stress is applied and the resultant sinusoidal strain is measured

9.2.3 Measuring Geometries [5]

A range of measuring geometries[1] is available to the user, and the particular geometry chosen will depend on the characteristics of the material being tested and the rheological information that is required. A list of the more common measuring geometries is presented below; however, situations can arise in which specialized spindle geometries are necessary to achieve optimum results. Figure 9.6 gives some considerations when choosing a measuring system.

- Disc spindles (Figure 9.7(a))—versatile general-purpose spindles. Disc spindles produce accurate, reproducible apparent viscosity determinations in a wide variety of liquids
- Cylindrical spindles (Figure 9.7(b))—scientifically defined spindle geometry for calculating shear stress and shear rate values as well as viscosity. Operating parameters are similar to those of disc spindles. Cylindrical spindles are particularly valuable when measuring non-Newtonian fluids
- Vane spindles (Figure 9.7(c))—used for measuring the viscosity of paste-like materials, gels, and fluids in which suspended solids migrate away from the measurement surface of standard spindles

---

[1] "More solutions to sticky problems" by Brookfield Engineering Laboratories Inc. Ref. [5] has been used during the compilation of this chapter.

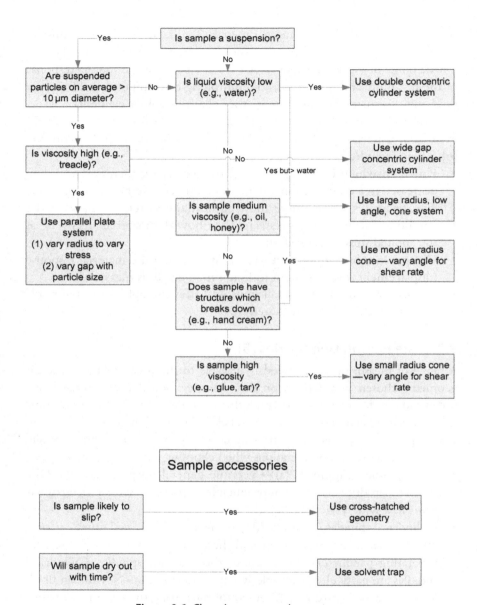

Figure 9.6 Choosing a measuring system.

- T-bar spindles (Figure 9.7(d))—generally used in conjunction with a Helipath accessory, T-bar spindles are used for measuring the viscosity of nonflowing or slow-flowing materials such as pastes, gels, and creams. The Helipath accessory raises and lowers the viscometer, while the T-bar

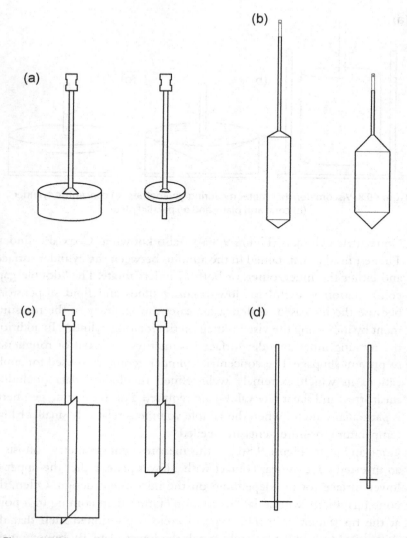

Figure 9.7 Viscometer spindles. (a) disc spindles; (b) cylindrical spindles; (c) vane spindles; and (d) T-bar spindles.

spindle rotates in the sample material causing the crossbar of the spindle to continuously cut into fresh material, thereby describing a helical path through the sample as it rotates. By comparison, conventional rotating spindles tend to form channels by pushing the sample material aside, resulting in a continuously decreasing viscometer reading

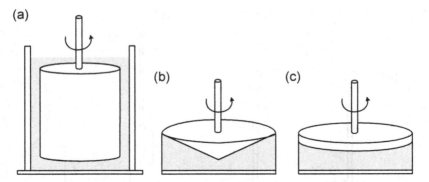

Figure 9.8 Viscometer/rheometer mearuring geometries. (a) concentric cylinders; (b) cone and plate; and (c) parallel plate.

- Concentric cylinders (Figure 9.8(a))—also known as Coaxial cylinders. The test fluid is maintained in the annulus between the cylinder surfaces and either the inner, outer, or both cylinders rotate. The "double gap" configuration is useful for low-viscosity fluids and fluid suspensions, because the increased total area increases the accuracy of the measurement by increasing the viscous drag on the rotating cylinder. In addition, the rotating inner cylinder surface is sometimes serrated or roughened to prevent slippage. The concentric cylinder geometry is used for applications in which extremely well-defined rheological data (including shear stress and shear rate values) are required. This measuring geometry is particularly useful when the sample volume is relatively small, or high temperature measurements are needed
- Cone and plate (Figure 9.8(b))—this measurement geometry consists of an inverted cone in near contact with a lower plate. Either the upper or lower surface rotates, depending on the instrument design. Often, the cone terminates with a flat "truncation" rather than coming to a point at the tip (Figure 9.9). This type of cone is positioned such that the theoretical (missing) tip would touch the lower plate. By removing the tip of the cone, a more robust measuring system is produced.

The cone and plate geometry is used for the accurate determination of absolute viscosity, as well as for advanced rheological flow analysis of non-Newtonian fluids (e.g., highly viscous pastes, gels, and concentrated suspensions). However, because strain and shear rate are calculated using the angular displacement and the cone/plate gap distance, the correct positioning of the cone (often referred to as "gap setting") is critical to the accuracy of the results.

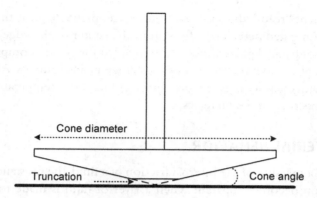

Figure 9.9 Cone and plate geometry.

A cone and plate geometry is not recommended when performing temperature sweeps or step changes unless the rheometer is fitted with an automatic system for thermal expansion compensation. It is only recommended if the sample contains particulate material in which the mean particle diameter is 5–10 times smaller than the gap, because the particles can "jam" at the cone apex/truncation resulting in noisy data. Materials with a high concentration of solids are also prone to being expelled from the gap under high shear rates.

When compared to concentric cylinders, the cone and plate can also be used with very small sample volumes. However, with concentric cylinders, the shear rate varies across the measuring annulus. With the cone and plate, shear conditions are constant across the geometry.

- Parallel plate (Figure 9.8(c))—the parallel plate geometry can be considered a simplified version of the cone and plate, having an angle of 0°. The test fluid is constrained in the narrow gap between the two surfaces. As with the cone and plate, the parallel plate geometry is most often used for highly viscous pastes, gels, and concentrated suspensions. The parallel plate (or plate–plate) system, like the cone and plate, requires a small sample volume. It is not as sensitive to the gap setting, because it is used with a separation between the plates measured in mm. Because of this, it is ideally suited for testing samples over a range of temperatures, or for making measurements on particulate material containing larger particles.

The main disadvantage of parallel plates comes from the fact that the shear rate produced varies across the sample. In most cases, the instrument software takes an average value for the shear rate. However, different instruments may produce different results depending on where each instrument "reads" the shear rate on the geometry (e.g., at the edge of the rotating geometry, or halfway from the center to the edge). As such, the use of parallel plates is not recommended for critical comparisons of test samples at defined shear rates on different instruments. Also, when using wider gap settings, the possibility of forming a temperature gradient across the sample increases.

9.3 MATERIAL BEHAVIOR

Viscosity is a measure of the internal friction of a fluid, the resistance to flow. This friction becomes apparent when a layer of fluid is made to move in relation to another layer. The greater the friction, the greater the amount of force required to cause this movement, which is known as "shear." Shearing occurs whenever the fluid is physically moved or disturbed, as in pouring, spreading, spraying, mixing, and so on. Highly viscous fluids require more force to move than do less viscous materials. Examples of shear rates that formulations experience during some typical processes and applications are listed in Tables 9.1 and 9.2 [7,8]. This section explains how materials are classified in relation to how that they behave when exposed to shear, and what things can affect rheological measurements [5].

9.3.1 Newtonian Fluids

For Newtonian fluids, the increase in shear stress is directly proportional to increasing shear rate (Figure 9.2). This means that, at a given temperature, the viscosity of a Newtonian fluid remains constant as the shear rate is varied (Figure 9.1).

Table 9.1 Process shear rates

Process	Typical shear rate range (s^{-1})
Sedimentation	10^{-3} to 10^{-1}
Sagging	10^{-2} to 10^{-1}
Leveling	10^{-2} to 10^{-1}
Extruding	10^{0} to 10^{2}
Pumping	10^{0} to 10^{3}
Chewing	10^{1} to 10^{2}
Brushing	10^{1} to 10^{2}
Stirring	10^{1} to 10^{3}
Mixing	10^{1} to 10^{3}
Curtain coating	10^{2} to 10^{3}
Rubbing	10^{4} to 10^{5}
Spraying	10^{4} to 10^{5}

9.3.2 Non-Newtonian Fluids

Emulsions, suspensions, polymer solutions, and gels are all examples of non-Newtonian fluids—that is, their viscosity is not a fixed value but changes as the shear rate is varied. The measured viscosity for non-Newtonian fluids is called the "apparent viscosity" due to its dependence on the shear rate. Several types of non-Newtonian behavior are possible, characterized by the way that the fluid's viscosity changes in response to changes in shear rate (Figure 9.2).

- Pseudoplastic—this type of fluid will display a decreasing viscosity with increasing shear rate (Figure 9.10), and is the most common form of non-Newtonian behavior. This type of flow behavior is also called shear thinning. Some common shear-thinning fluids include paints, emulsions, and dispersions. Shear-thinning products are formulated such that they exhibit good suspension stability or drip resistance when at rest, but thin down when "worked" to make application easy (e.g., rubbing in a hand cream, applying paint to a wall). For shear-thinning products, measuring viscosity at a single shear rate will only provide a tiny glimpse of

Table 9.2 Application shear rates

Application	Typical shear rate range (s⁻¹)
Suspending ingredients	10^{-3} to 10^{-1}
Pouring	10^{1} to 10^{2}
Extrusion from a tube	10^{2}
Topical application of creams and lotions	10^{2} to 10^{4}
Applying lipstick	10^{3} to 10^{4}
Brushing nail polish	10^{3} to 10^{4}

Reproduced with permission of Taylor and Francis Group LLC Books, from "Rheological Properties of Cosmetics and Toiletries", Dennis Laba (Ed), Marcel Dekker Inc, New York 1993; permission conveyed through Copyright Clearance Center Inc.

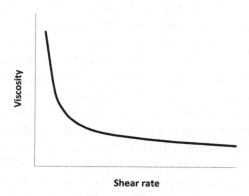

Figure 9.10 Shear-thinning (pseudoplastic) behavior.

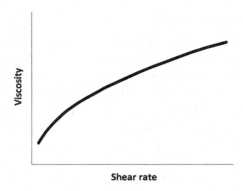

Figure 9.11 Shear-thickening (dilatant) behavior.

Figure 9.12 Plastic behavior.

the full picture. Instead, a flow curve of viscosity across a range of shear rates is far more meaningful, and will enable the viscosity value to be determined at a shear rate relevant to the process or product usage conditions.

- Dilatant—this type of fluid will display an increasing viscosity with increasing shear rate (Figure 9.11). This type of flow behavior is also called shear thickening, and is less common than pseudoplasticity. Shear thickening is frequently seen in fluids containing high levels of deflocculated solids, such as clay slurries (e.g., bentonite), candy compounds, corn starch/water mixtures, and sand/water mixtures.
- Plastic—this type of fluid will behave as a solid under static conditions. A certain amount of force must be applied to the fluid before any flow is induced. This force is called the "yield stress." Once the yield value is exceeded and flow begins, plastic fluids may display Newtonian, pseudoplastic, or dilatant flow characteristics (Figure 9.12). Plastic behavior is exhibited by tomato ketchup. The yield value of tomato ketchup will

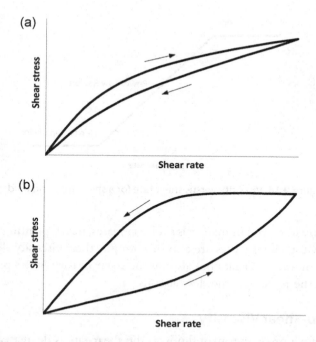

Figure 9.13 Flow curves/hysteresis loops. (a) thixotropic material and (b) rheopectic material.

often prevent it from pouring from the bottle until the bottle is shaken or struck, allowing the ketchup to flow freely.

- Thixotropic and rheopectic—these fluids display a change in viscosity with time when exposed to conditions of constant shear rate. Thixotropic fluids exhibit a decrease in viscosity; rheopectic fluids exhibit an increase in viscosity.

Also, when the shear rate is increased to a certain value, then immediately decreased to the starting point, thixotropic and rheopectic fluids behave as shown in Figure 9.13. The shear rate "up" and "down" curves do not coincide but form a "hysteresis loop," caused by a decrease in the fluid's viscosity with increasing time of shearing. Some thixotropic fluids, if allowed to stand undisturbed for a while, will regain their initial viscosity, whereas others never will.

Both thixotropy and rheopexy may occur in combination with any of the previously discussed flow behaviors, or only at certain shear rates. Under conditions of constant shear, some fluids will reach their final viscosity value in a few seconds, whereas others may take up to several days. Rheopectic fluids are rarely encountered. Thixotropy, however, is

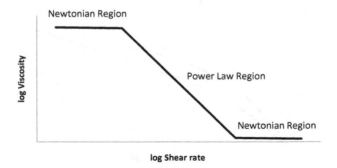

Figure 9.14 Viscosity versus shear rate for a shear-thinning fluid.

frequently observed in materials such as greases, heavy printing inks, and paints. Most solder pastes are also thixotropic; their viscosity depending not just on the shear rate but also on the shear history of the paste. After stirring, the paste becomes less viscous.

9.3.3 Zero-Shear Viscosity

In general, for non–Newtonian fluids, as the shear rate is decreased the viscosity increases. This increase does not continue indefinitely, but reaches a plateau (Figure 9.14). Zero-shear viscosity is effectively the viscosity of a product while at rest. The zero-shear viscosity of a suspending agent is key to the stability of a suspension or emulsion, as this is related to the settling or creaming rates of particles or droplets.

The zero-shear viscosity of thickeners and gelling agents can be used to compare between different batches of these ingredients. It is also a sensitive indicator of stability changes in a product, or changes in the formulation or manufacturing process.

9.3.4 Yield Stress Testing [9]

The yield stress (σ_y) is the stress that must be applied to disrupt the structure in a material and cause it to flow. The thickening effects of that structure are lost, viscosity decreases markedly, and the material makes the transition from an apparent solid to a free-flowing liquid (Figure 9.15).

Yield-stress measurements can be used to help interpret many aspects of material processing, handling, storage, and performance properties—for example, how creams and ointments feel as they are being squeezed out of a tube; the ability of a fluid to hold particles in suspension.

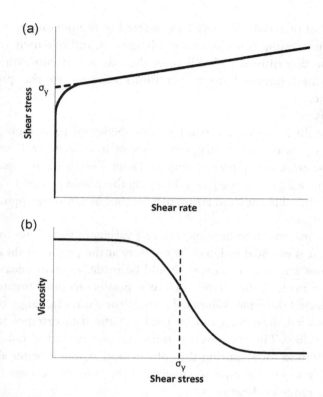

Figure 9.15 Yield stress (σ_y). (a) shear stress as a function of shear rate and (b) viscosity as a function of shear stress.

9.3.5 Factors Affecting Rheological Measurements

This section looks at some of the factors that can affect or influence rheological measurements. When developing rheological tests, these factors should be specified as part of the instrument setup and measurement conditions.

- Temperature

One of the most obvious factors that can have an effect on the rheological behavior of a material is temperature. Viscosity typically, but not always, exhibits an inverse relationship with temperature—one exception is in the case of thermosetting gels. The extent to which temperature affects the recorded viscosity varies from material to material. To compare viscosity measurements (e.g., for stability assessment of formulations, for comparing formulation prototypes, or for assessing different ingredients), it is important to note the temperature at which the viscosity measurement is made. Also, the effect of temperature on viscosity is essential in the evaluation of materials that will be subjected to temperature variations in use or on processing (such as motor oils, greases, and hot-melt adhesives), and can be determined by evaluating the viscosity–temperature profile under defined imposed-shear conditions relevant to the process or product use.

- Shear Rate

As most fluids in the real world are non-Newtonian, an appreciation of the effect of shear rate on material behavior is a necessity. It would, for example, be disastrous to try to pump a dilatant fluid through a system, only to have it go solid inside the pump, bringing the whole process to an abrupt halt. Similarly, a dilatant hand cream may be too thick during application for its intended use.

When a material is to be subjected to a variety of shear rates in processing or use, it is essential to know its viscosity at the projected shear rates. If these are not known, an estimate should be made. Viscosity measurements can then be made at shear rates as close as possible to the estimated values. If the projected shear rate values fall outside the shear rate range of the viscometer, several shear rates can be used and the data extrapolated to the projected values. This, however, is time consuming and is not the most accurate method for acquiring this information. A much better alternative is to use a rotational rheometer to record the flow curve over the entire anticipated range of shear rates.

Examples of materials that are subjected to, and are affected by, wide variations in shear rate during processing and use are: paints, cosmetics, liquid latex, coatings, and certain food products. Tables 9.1 and 9.2 show typical examples of varying shear rates.

- Measuring Conditions

Variables such as the model of the viscometer or rheometer, spindle/speed combination, sample container size, sample temperature, entrained air within the sample, and sample preparation technique can all affect the viscosity measurement.

Another factor that may affect the viscosity measurement is the homogeneity of the sample. More consistent results are generally obtained from homogeneous samples. However, mixing or shaking the sample to obtain homogeneity may affect the measured viscosity.

- Time

Time dependence includes those effects associated with transient flow conditions as well as those effects associated with irreversible changes that result from shear history.

How long the sample is sheared before taking a reading will affect the results for time-dependent materials (thixotropic and rheopectic samples).
• Previous History
Thixotropic materials are particularly sensitive to prior history. Their viscosity will be affected by stirring, mixing, pouring, filling, coating, or any other activity that produces shear in the sample. Storage conditions and sample preparation techniques must be designed to minimize their effect on subsequent viscosity tests; for example, sample preconditioning instructions (premixing or rest periods).

9.3.6 Characteristics of Dispersions and Emulsions [5]

Dispersions and emulsions are multiphase materials, and may consist of one or more solid phases dispersed in a liquid phase. One characteristic of these systems that can play a big role in their rheology is the state of aggregation of the solid-phase material: whether the particles are separate and distinct, or are clumped together (flocs); the size of the flocs; how tightly they are stuck together.

• Volume—If the flocs occupy a large volume, the measured viscosity will tend to be higher than if the floc volume was smaller. This is due to the greater force required to dissipate the solid component
• Aggregation—If the flocs are aggregated, the reaction of the aggregates to shear can result in shear-thinning (pseudoplastic) flow. At low shear rates, the aggregates may be deformed but remain essentially intact. At higher shear rates, the aggregates might be broken down into individual flocs, decreasing friction, and therefore viscosity
• Internal Bonding—If the bonds within the aggregates are strong, the system may display a yield value
• Time—The flocculated structure may be destroyed with time as the material is sheared. If the shear is subsequently decreased and the rate of relinking of the flocs is low, thixotropic behavior will be observed.

The use of flocculating or deflocculating agents can modify the attraction between particles in the dispersed phase, and thereby help control the rheology of the system. The shape of the particles also contributes to the system's rheology. As the particles are constantly being rotated in the flowing medium, spherical particles will be able to rotate more freely than rod-like particles. And finally, the stability of the dispersed phase is critical to the viscosity of a multiphase system. If the dispersed phase has a tendency to settle with time, the fluid will become nonhomogeneous and the rheological characteristics of the system will change—typically, the viscosity will decrease.

9.4 RHEOLOGICAL STUDIES

This section outlines some of the more common rheological studies that are performed using rotational viscometers and rheometers.

9.4.1 Viscosity

Viscosity testing entails applying a shear stress to a fluid, and then measuring the rate of material flow caused by this stress. For viscoelastic materials, the resulting viscosity varies with the applied stress—consequently, for rotational instruments, the measuring geometry (e.g., spindle) and speed of rotation will affect the viscosity reading. The viscous drag, or resistance to flow, will increase as the spindle size and/or rotational speed increases.

- Selecting the Spindle and Speed

  When developing a viscosity method using a rotational viscometer, the goal is to obtain a viscometer dial or display reading of approximately half of the full-scale deflection or range. This target is chosen such that when conducting multiple tests, any changes in the viscosity (caused by, e.g., different batches of product, or when carrying out stability assessments over time) can be captured using the same spindle and speed combination. Selecting the spindle and speed combination is typically done by trial and error.

  For Brookfield® viscometers, dial or display readings between 10 and 100 are recommended [5]. This is because the readings are accurate to within ±1% of the full-scale range and, although the accuracy improves as the reading approaches 100, it can become significant for readings below 10. If the reading is over 100, a slower speed and/or smaller spindle should be used. Conversely, if the reading is under 10, a higher speed and/or larger spindle should be used.

- Sample Preparation

  The sample fluid should be free from entrapped air and kept at a constant and uniform temperature. The correct size sample and container should also be used. A sample size of 600 ml in an 83-mm internal diameter container is recommended for Brookfield® viscometers. Small containers can result in an increase in viscosity readings. If the same container is used for subsequent tests, however, the viscosity results relative to each other will still be valid. The effect of shear must also be taken into consideration when preparing the sample. For example, if the material is shear thinning and is contained in a tube, squeezing the sample out of the tube will reduce the viscosity. A product like this might need to stand undisturbed for a period of time after being squeezed from the tube prior to commencing the viscosity measurement.

9.4.2 Flow

In flow experiments, the shear stress is increased then decreased and the resultant shear rate measured, producing a flow curve (Figure 9.13). If the "up" and "down" curves do not superimpose, a hysteresis loop is formed showing that the material is either thixotropic or dilatant. Flow experiments cause the material structure to break, generating information regarding the viscosity and yield stress of the fluid.

Figure 9.14 shows a plot of log viscosity against log shear rate, which is representative of most non-Newtonian materials. At low shear rates, a region exists that is known as the "first Newtonian plateau" (the zero-shear viscosity). The viscosity is constant in this region because no structure is being destroyed. As the shear rate is increased, structure is destroyed by the shearing action leading to a sharp decrease in viscosity (the Power Law region). Finally, when no structure remains to destroy, the viscosity becomes constant at the "second Newtonian plateau" (infinite-shear viscosity). Mathematical models [3,10] in the instrument software are used to analyze the shear stress and shear rate data, and thereby help characterize the non–Newtonian behavior of the fluid—for example:

- Bingham—used for describing viscoplastic fluids that exhibit a yield response. The classic Bingham material is an elastic solid at low shear stress and a Newtonian fluid above a critical value (the "Bingham yield stress"). The plastic viscosity region exhibits a linear relationship between shear stress and shear rate, with a constant viscosity equal to the plastic viscosity (Figure 9.2).

The Bingham model is simple and is used to describe the behavior of concentrated suspensions of solid particles in Newtonian liquids. The Bingham model can be written mathematically as:

$$\sigma = \sigma_0 + \eta_B \dot{\gamma}$$

in which σ_0 is the yield stress and η_B is the Bingham viscosity or plastic viscosity. It should be noted that the Bingham viscosity is not a real viscosity value; it just describes the slope of the Newtonian portion of the curve.

- Casson—used for describing flow in viscoplastic fluids that exhibit a yield response. The Casson model is an alternative model to the Bingham model, and has all of the components in the Bingham equation raised to the power of 0.5, but has a more gradual transition between the yield and Newtonian regions. The Casson equation can be written as:

$$\sigma^{0.5} = \sigma_0^{0.5} + \eta_C^{0.5}\dot{\gamma}^{0.5}$$

in which η_C is the Casson viscosity, which relates to the high shear rate viscosity.

- Power Law—used for describing either pseudoplastic or shear-thickening behavior in materials that show a negligible yield response and a varying viscosity. The Power Law can be written as:

$$\sigma = K\dot{\gamma}^n$$

in which K is the consistency and n is the shear-thinning index (Power Law exponent). A log–log plot of σ versus $\dot{\gamma}$ gives a slope n; where $n < 1$ for shear-thinning materials and $n > 1$ for shear-thickening materials.

- Herschel–Bulkley—used to describe viscoplastic materials that exhibit a yield response with a Power Law relationship between shear stress and shear rate above the yield stress. The Herschel–Bulkley equation is basically a Power Law model with a yield stress term, and is written as:

$$\sigma = \sigma_0 + K\dot{\gamma}^n$$

- Sisko—is used to describe flow in the Power Law and upper Newtonian regions, and is written as:

$$\eta = \eta_\infty + K\dot{\gamma}^n$$

in which η_∞ is the viscosity at infinite shear rates.

- Cross—is a more complex (four parameter) model used to describe pseudoplastic flow covering the entire shear rate range, and no yield stress. The Cross equation can be written as:

$$(\eta-\eta_\infty)/(\eta_0-\eta_\infty) = 1/(1 + (\lambda\dot{\gamma})^m)$$

in which η_0 is the viscosity at zero shear rates, λ is a time constant, and m is a dimensionless exponent.

9.4.3 Creep [11,12]

In creep experiments, a small constant stress is applied to a sample and the resulting elastic deformation and/or viscous flow over medium to long periods is monitored (Figure 9.16). When a viscoelastic material is subjected to a creep test the initial stage of the test is dominated by elastic, recoverable deformation. As the test progresses the sample exhibits viscoelastic deformation, eventually resulting in nonrecoverable viscous (steady state) flow.

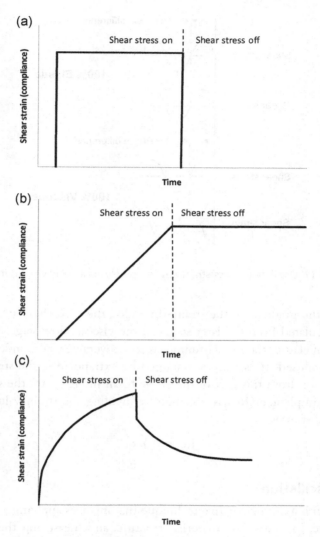

Figure 9.16 Creep curves. (a) pure elastic; (b) pure viscous; and (c) viscoelastic.

If required, the stress can then be removed and the relaxation of the sample measured. Newtonian samples exhibit no recovery at all, whilst instantaneous recovery of the total deformation will be observed with purely elastic materials. Viscoelastic samples will only recover a portion of the deformation.

Creep experiments are performed on a controlled-stress rheometer by distorting the material structure at low stresses (within the linear viscoelastic range of the material).

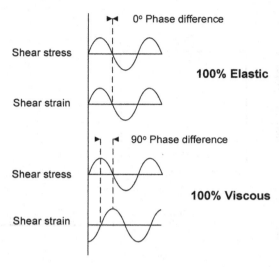

Figure 9.17 Oscillation stress/strain waves—purely elastic and viscous materials.

- From the gradient of the strain–time plot, the zero-shear viscosity can be calculated from the later stages of the viscous-flow stage. The equilibrium elastic strain (maximum elastic recoverable strain under the specific imposed stress) can be obtained by extrapolating the straight-line regression from this part of the curve to an intercept on the strain axis. The compliance ($J(t)$) is obtained by dividing the strain values by the applied stresses:

$$J(t) = \gamma(t)/\sigma_0$$

9.4.4 Oscillation

Oscillation is a nondestructive technique that involves applying a sinusoidal stress wave, to distort the material structure, and measuring the resulting strain wave. If a material is purely elastic (Figure 9.17), the phase difference between the stress and strain waves is 0° (i.e., the stress and strain waves are in phase). If a material is purely viscous, the phase difference is 90° (i.e., the stress and strain are out of phase). Most materials are viscoelastic, and therefore have a phase difference between these two extremes (Figure 9.18). The phase difference together with the amplitudes of the stress and strain waves provide information relating to the viscous and elastic contributions to the structure of the material. Oscillation rheology testing is almost always performed at very low applied stresses and strains, often significantly below the

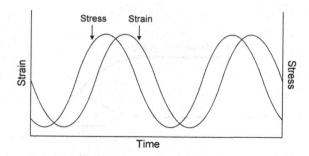

Figure 9.18 Oscillation stress/strain waves—viscoelastic material.

yield point of a sample (within the linear viscoelastic range of the material). Like creep, oscillation also provides information on viscoelastic behavior but is far more suitable for measurements over short timescales (0.25–1000s).

Several types of oscillation experiments are used, including torque (stress) sweep, frequency sweep, time sweep, and temperature sweep. Typically measured parameters include: complex modulus (G^*), elastic (or storage) modulus (G'), viscous (or loss) modulus (G''), phase angle (δ), and tangent of the phase angle ($\tan \delta$).

- Complex modulus (G^*) is a measure of the resistance to deformation of the sample.
- Elastic (or storage) modulus (G') is a measure of the energy that is stored in a material in which a deformation has been imposed. The storage modulus is that proportion of the total rigidity (the complex modulus) of a material that is attributable to elastic deformation.

$$G' = G^* \cos \delta$$

- Viscous (or loss) modulus (G'') is a measure of the energy that is dissipated in a material on which deformation has been imposed. The loss modulus is that proportion of the total rigidity (the complex modulus) of a material that is attributable to viscous flow, rather than elastic deformation.

$$G'' = G^* \sin \delta$$

- Phase angle (δ) is the phase difference between the stress and strain curves in an oscillatory test, and is a measure of the extent of elastic (or "gel") structure within a material—a low value indicating greater elastic behavior. A drop in complex modulus during the yielding process is matched by an increase in phase angle, indicating that there is a breakup of internal elastic structure in the sample as the imposed stress is increased.

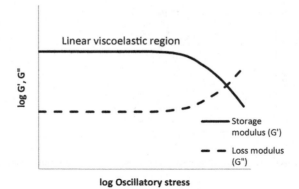

Figure 9.19 Oscillatory stress sweep showing storage (G′) and loss (G″) moduli.

- The tangent of the phase angle (tan δ) is the ratio of the viscous modulus (G″) to the elastic modulus (G′), and is a useful quantifier of the presence and extent of elasticity in a fluid. Tan δ values of less than unity indicate elastic–dominant (i.e., solid–like) behavior, and values greater than unity indicate viscous–dominant (i.e., liquid–like) behavior.

$$\tan \delta = \frac{G''}{G'}$$

- Oscillation Stress Sweep
  In an oscillation stress sweep test, the sample is subjected to small-amplitude oscillatory (i.e., clockwise, then counterclockwise) shear. In the early stages of the test, the stress is sufficiently low to preserve structure. This plateau region (the linear viscoelastic region) provides information relating to the resistance of the material to deformation (the stiffness of the product). As the test progresses, the increasing applied stress causes the ultimate disruption of structure (the product yields) and is seen as a decrease in elasticity (storage modulus, G′) and rigidity (complex modulus, G*), and an increase in the loss modulus (G″)—Figure 9.19. Yield stress is a useful practical measure of the stress required to induce flow in a product. In fact, when exposed to stresses below the yield stress, viscoelastic materials such as creams, lotions, and ointments do flow, but at a very low rate (creeping flow). The point at which the elastic and viscous moduli cross marks the transition from the "solid" plateau (elastic) region to the "fluid" (viscous) region.
- Oscillation Frequency Sweep
  Oscillation frequency sweep tests enable the viscoelastic properties of a sample to be determined as a function of timescale [9,13]. In the test, the

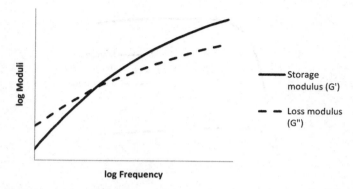

Figure 9.20 Oscillation frequency sweep for a typical polymer solution.

structural response of the sample to small-deformation oscillations is assessed over a range of frequencies. For analysis, the storage and loss moduli are plotted against the frequency. The modulus (G' or G'') that is dominant at a particular frequency gives an indication as to whether the fully structured material appears to be elastic or viscous, in a process of similar timescale. If the elastic and viscous moduli cross over, this frequency (the crossover frequency) marks the transition from viscous to elastic behavior. Figure 9.20 shows a frequency sweep for a typical polymer solution—the material is mainly viscous at low frequencies, and elastic at higher frequencies.

• Time Sweep
Some materials, for example, dispersions (suspensions, emulsions, foams) and polymers, undergo structural changes with time—including polymer degradation, solvent evaporation, dispersion settling, polymer curing, and time-dependent thixotropy. Oscillatory time sweeps provide information about how the structure within a material changes with time, without damaging that structure. Additionally, data collection in rheological tests such as creep/recovery, flow, oscillatory stress, and frequency sweeps should not begin while network rearrangements are occurring in the dispersed phase of the test material—for example, while the sample is recovering after being loaded onto the rheometer. This requires that the time-dependent structural effects, due primarily to the microscopic interactions of the dispersed phase within the liquid medium and itself, are first tested [14]. Figure 9.21 shows an example of changing storage modulus with time.

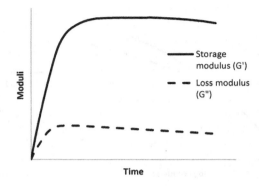

Figure 9.21 Changing storage (G′) and loss moduli (G″) with time.

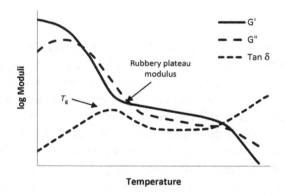

Figure 9.22 Typical temperature sweep curve for a polymeric material.

- Temperature Sweep
  Temperature can drastically affect how a material responds to applied
  stress. Most materials are solid-like when cold and become more liquid-
  like at higher temperatures. In contrast, however, some materials exhibit
  the opposite effect—for example, some proteins begin to denature and
  cross-link on heating, forming a gel. Also, some temperature changes are
  reversible, whereas others are not.
  During a temperature sweep test, a sinusoidal strain is applied at a constant frequency, and successive measurements are taken as the temperature is automatically increased or decreased from selected lower or
  upper temperature limits. This test is often used to analyze storage
  modulus (G′), loss modulus (G″), and Tan δ as a function of temperature (Figure 9.22), from which information can be obtained to characterize polymeric materials, such as molecular weight, molecular

weight distribution, glass transition temperature (T_g—the point at which the structure changes due to temperature), melting point (T_m), rubbery plateau modulus, effect of cross-linker and crystallization, and compatibility of polymer blends [15].

9.5 PRACTICAL APPLICATION OF RHEOLOGICAL DATA

Suspensions, colloidal dispersions, and emulsions are complex formulations comprising solvents, suspended particles of varying size and shape, and other additives. Many factors affect formulation stability—for example: hydrodynamic forces, Brownian motion, strength of interparticle interactions, volume fraction, electrostatic forces, steric repulsion, and size and shape of particles. Rheology-based measurements can help predict which formulations might exhibit flocculation, coagulation, or coalescence, resulting in undesired effects such as settling, creaming, separation, and so forth. They can also provide critical information for product and process performance in many industrial applications.

This section presents examples of how typical rheological data generated in the types of studies listed in Section 9.4 are used and interpreted to characterize semisolid formulations and their behavior in the real world [16].

- Time Sweep

  When a sample is applied to the rheometer plate—either by squeezing it from a tube, scooping it out of a container, and so on—the structure within the sample will be disrupted. For this not to affect the subsequent rheological measurements, this structure needs to be allowed to rebuild before starting the test. Running a time sweep will show how quickly this structure is rebuilt (Figure 9.21), and will give an indication of the "resting time" needed before commencing testing.

  Time sweeps are also used to monitor how quickly the material will recover structure after various stages in processing—for example, after filling a cream into a tube. This information is useful when processing suspensions in which the suspended material may sediment out if the structure builds too slowly after mixing is stopped or after filling into the final product container.

  The time sweep in Figure 9.23 shows a sample that is liquid-like initially, the viscous modulus (G'') being higher than the storage modulus (G'). At the crossover point, the formulation gels, the structure becomes predominantly elastic, and the sample becomes more resistant to flow. This type of information would be useful for a formulation that needed to be

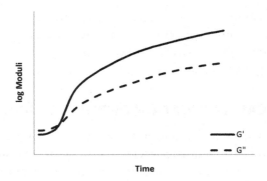

Figure 9.23 Time sweep for dental molding system.

worked or molded during application, before setting up as a final solid (e.g., the curing process in thermosets, two-part epoxy adhesives, and dental molds).

- Flow Curve
  Flow measurements are used to predict how semisolids and liquids will behave in real-life situations, such as pumping, stirring, and extrusion through a nozzle. The flow curve/hysteresis loop (Figure 9.13) gives information on sample breakdown and recovery—whether the sample structure will recover instantaneously when the stress is removed, or whether it will recover over a period of time. The zero-shear viscosity is related to the settling or creaming rates of particles or droplets, and can be used to determine the stability of suspensions and emulsions (Figure 9.14). Comparisons can be made between different batches and types of thickeners and gelling agents. It is also a sensitive indicator of stability changes in a product, or changes in the formulation or manufacturing process. The yield stress gives an indication of how easily flow can be induced in the product. The shape of the flow curve, and the presence of any hysteresis, will classify the flow behavior of the product—whether Newtonian, plastic, pseudoplastic, dilatant, thixotropic, and so forth (Figure 9.2).
  Flow behavior also gives an indication of the degree of dispersion and/or flocculation within the formulation. Dispersed systems have a higher viscosity than the pure continuous phase, and the viscosity increases when more dispersed particles are present. Flocculation leads to an increase in the effective volume fraction (due to the water trapped within the floc), and a corresponding increase in viscosity.

Ideally, for a pharmaceutical ointment or gel in a tube, the product should have a relatively high zero-shear viscosity (to prevent settling or sedimentation in the tube on storage). In use, the formulation structure should disrupt easily when the tube is squeezed (have low yield stress), flow readily when being applied to the skin (to improve skin feel and reduce drag on sensitive or broken skin), and then rebuild structure quickly (have negligible hysteresis) to prevent runoff from the skin before a dressing can be applied. Similar product characteristics are also favorable for processing the formulation in so much as the product should flow readily during pumping, stirring, and filling through a nozzle, and then rebuild structure rapidly when in the final packaging (e.g., tube) to prevent sedimentation of any dispersed ingredients.

- Stress Sweep

With semisolid formulations, viscoelastic behavior is related to the structure within the formulation. Oscillation stress sweeps can be used to evaluate formulation differences by performing experiments within the linear viscoelastic region—in which the properties being measured are independent of the imposed stress and strain levels. Stress sweeps are used to characterize and quantify the presence, rigidity, and integrity of a material's internal structure resulting from, for example, flocculation and interaction of dispersed particles or droplets, or cross-linking and entanglement of dissolved polymers.

The length of the linear viscoelastic region, and the values of the complex modulus (G^*) and phase angle (δ) provide information relating to the structure and stability of a formulation—a longer linear viscoelastic region generally indicates more stability, a higher value of G^* indicates more resistance to deformation, and a lower value of δ indicates more structure. Figure 9.24 shows how oscillation stress sweeps reveal significant differences in the behavior of a lotion, a cream, and an ointment [17]. As the oscillatory stress is increased, the samples maintain their resistance to deformation until they reach a critical range of stresses (the yield stress), whereupon they undergo a rapid transition from a high to low modulus. The plateau values for G^* provide information relating to the stiffness of the product.

Structured fluids, such as emulsions and suspensions, possess a yield stress—the level of applied stress required to induce flow in a product. The presence of a yield stress can impart desirable handling, appearance, and storage properties to a product. Below the yield stress, formulations may exhibit a slow creeping flow. The yield stress is crucial for

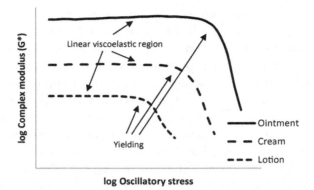

Figure 9.24 Oscillation stress sweep showing complex modulus (G*) for different product types.

determining not only the shelf life of a formulation, but also the ease of application for the end user—for example, it contributes to the skin feel of a cream during application. The curves in Figure 9.24 reveal a low relative yield stress for a lotion (a prerequisite for pourability) and a significantly higher relative yield stress for an ointment (a prerequisite for wash resistance). Yield-stress measurements can also be used to help interpret many aspects of material processing and performance—for example, pumping, mixing, and filling is easier for lower yield-stress formulations.

In the case of a gel, the length of the linear viscoelastic region provides a good indication of the resistance of the formulation to syneresis—the separation of liquid from the bulk formulation (seen as a liquid layer on the surface of the product).

The comparative length of the linear viscoelastic region, values of the moduli (G*, G', G''), the phase angle, and yield stress can be used to compare the effectiveness of different structure-building ingredients (e.g., thickeners) or the effect of other additives (either thickening or thinning) during formulation development exercises (Figure 9.25). Furthermore, monitoring changes in any of these parameters on storage of a product gives valuable information relating to formulation stability.

• Frequency Sweep

A frequency sweep allows the structural nature of the sample to be determined—for example, to distinguish between a particle solution, an entangled solution (e.g., a paste), or a three-dimensional network (e.g., a gel). The measurements need to be made within the linear viscoelastic

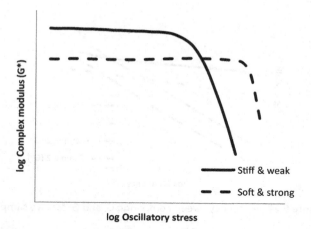

Figure 9.25 Oscillation stress sweep comparing the behavior of different formulations.

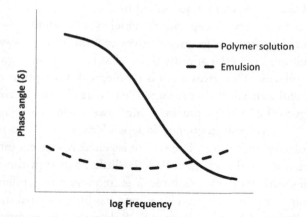

Figure 9.26 Frequency sweep showing differences between types of product.

region, obtained from the stress sweep. Figure 9.26 shows how the technique can differentiate between the "relaxable" structure found in a polymer solution (in which polymers disentangle to dissipate stored stresses) and the more permanent elasticity found in a flocculated suspension (emulsion system)—the lower value of the phase angle (δ) indicating a more "elastic" product. The presence of suspended solids, high additive concentrations, colloidal thickeners, and so on will induce some sort of structure upon the bulk phase. Information relating to the degree of dispersion and interparticle association can be seen from the profile

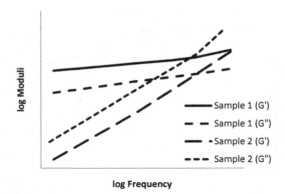

Figure 9.27 Frequency sweep comparing G′ and G″ for two samples.

of the frequency sweep data—the fully dispersed sample will give the most gelled system. Similarly, the effectiveness of surfactants and other colloidal thickeners can be quantified in situ.

With the frequency sweep curve, whichever modulus (G′ or G″) is dominant at a particular frequency gives an indication as to whether the fully structured material appears to be elastic or viscous in a process of similar timescale. This gives a good rheological description of how the product will behave during storage (i.e., "at rest") and during application. Figure 9.27 shows the frequency sweeps for two samples [18]. Sample 1 is relatively frequency independent, with G′ dominant over the entire frequency range. The system is gelled, has strong intermolecular associations, and is characterized by little change in the viscoelastic properties with frequency. Sample 2 is frequency dependent, with G″ being dominant throughout. This system has little internal structure and is easily disturbed. Similarly, Figure 9.28 shows the frequency sweeps for a viscoelastic solid and a gel. With the viscoelastic solid, the phase angle is low and G′ is dominant at low frequencies—indicating solid-like behavior. The phase angle increases and G″ becomes dominant as the frequency is increased. With the gel, the phase angle, G′ and G″ are frequency independent. A high phase angle signifies that any suspended particles can be deposited as sediment and settle if left long enough. In this case, the viscoelastic solid will show enhanced stability at rest (on storage), but the gel may show enhanced stability at higher frequencies (e.g., during transport).

Frequency sweeps can prove a useful tool when attempting to match textures and flow behaviors in thickened systems. In addition, when a pigment or thixotrope needs to be well dispersed into a fluid matrix, the

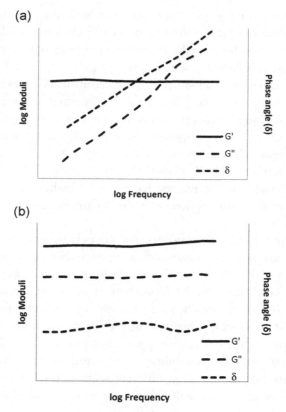

Figure 9.28 Frequency sweeps. (a) viscoelastic solid and (b) gel.

process can be optimized by sampling and running repeated frequency sweep tests.

- Temperature Sweep

 A temperature sweep enables information to be obtained relating to the amount of thinning that occurs when a formulation is heated. Many pharmaceutical and cosmetic products are stored at one temperature (e.g., 25 °C), but applied at body temperature (e.g., 37 °C). A lip balm, for example, needs to retain a relatively high G' and $\mathrm{Tan}\,\delta < 1$ as the temperature increases over this range such that it will be relatively stiff at room temperature, yet will adhere to the lips (not be runny) and provide a soft feel. Thermosets are processed at elevated temperatures to obtain the required level of moldability, as well as to subsequently complete the cure. Hence, manufacturers working with these materials are interested in the changes in viscosity that occur with temperature and time to determine the minimum viscosity achieved (which is an indication of the maximum level of moldability), the time of onset of gelation (which is the time available to complete molding), and the gel temperature (which is the temperature that must be achieved before gelation occurs) [6].

The holding power of hot-melt pressure-sensitive adhesives is determined by their resistance to shear [15]. Rheologically, a higher value of G', higher viscosity, and longer rubbery plateau modulus indicate higher shear resistance (Figure 9.22). A temperature sweep will also provide information on the adhesive's holding power at high temperatures. The crossover temperature, in which $G' = G''$ ($\text{Tan}\,\delta = 1$), can be correlated to the softening point of the adhesive. Adhesives that exhibit a higher crossover temperature tend to show higher shear adhesion failure temperatures.

- Creep Recovery

  Creep testing is used to investigate the viscoelastic structure of materials over medium and long timescales (e.g., on stability). A low shear stress that is well within the linear viscoelastic limit for the material under test is applied, and the response (compliance) of the material is measured (Figure 9.16(c)). The steady-state gradient of the strain–time plot can be used to obtain the zero-shear viscosity. The zero-shear viscosity will give an indication of the tendency for particulate materials to settle out from a suspension (suspension stability). Comparing the zero-shear viscosities of different formulations (e.g., different thickeners or suspending agents, different concentrations) will give a comparative ranking of their relative stability.

  Creep tests are also used to explore product behavior in which low stresses are maintained for a period—for example, sagging of coatings due to gravity, or leveling of paints due to surface tension. On a nonhorizontal surface, a coating will experience tangential shear stress due to gravity. Coatings that have a yield stress will not sag unless the gravitational shear stress exceeds the yield stress.

9.5.1 Reported Examples

This section summarizes a few examples of the practical use of rheological data.

- The application of rheological measurements to the design of a thickened disposal system for mineral waste tailings has been presented by Boger, Scales, and Sofra [19]. Here, rheological concepts covering each stage of the disposal process were used to select an appropriate thickener, including: yield and flow under low and high shear (to evaluate different pumping forces); and change in yield stress with solids content (to prevent settling on standing).

- Jiao and Burgess [20] have documented the use of rheological measurements to help assess the stability of water-in-oil-in-water (W/O/W) multiple emulsion systems. Multiple emulsions contain both W/O and O/W simple emulsions, and require the use of at least two surfactants to stabilize the system—one with a low hydrophilic–lipophilic balance (HLB) to stabilize the primary (W/O) emulsion, and one with a high HLB to stabilize the secondary (O/W) emulsion. The stability of the overall emulsion system is dependent on the relative proportion of each surfactant. Using measurements of the viscosity at low stress, as well as oscillatory stress sweeps, the stability of the W/O/W emulsions formulated with various surfactant ratios was assessed to determine the optimum formulation.

- Using full-thickness skin membranes taken from the abdominal skin of the hairless rat, the relationship between the *in vitro* penetration of a drug (indomethacin) has been compared to the rheological characteristics for several O/W ointments by Shon and Lee [21]. It was found that the absorption of the drug through the skin membrane had a close relationship to the spreadability of the ointment.

- In a presentation by Yu, Bunge, and Wilkin on the bioequivalence of topical products [22], the use of rheology was discussed as one means of determining structural similarity between formulations. Furthermore, to obtain similar spreadability, the need for the viscosity–shear rate curves and yield stress to be the same for the different formulations was highlighted.

- An application note produced by Bohlin Instruments Inc.[16] looks at the information generated by various rheological techniques (viscosity, dynamic oscillation, creep testing, and stress relaxation), and points out drawbacks for different molecular weight polymers. Due to the duration of the different techniques, relaxation experiments were considered the most useful as polymer QC tests. The information generated by the relaxation experiments was also considered to be useful in practical applications in which the plastic needs to relax before cooling, such as cycle-time settings for injection molding.

In summary, rheology testing can be very effective for screening different formulations during the development stage of projects, for competitive product analysis, or for providing information relating to formulation stability. It is R&D's job to characterize the flow properties of new formulations, which includes measuring physical properties like viscosity and yield stress. Routine QC test methods for batch-to-batch quality control, however, usually include a viscosity test (e.g., using a Brookfield® viscometer), but seldom one for yield stress.

REFERENCES

[1] Gupta P, Garg S. Recent advances in semisolid dosage forms for dermatological application. Pharm Technol March 2002;26:144–62.
[2] Whittingstall P. The importance of rheology. Am Lab Febru1y 1993:41–2.
[3] Hackley VA, Ferraris CF. Guide to rheological nomenclature: measurements in ceramic particulate systems. National Institute of Standards and Technology. NIST special publication 946; January 2001.
[4] US Pharmacopeia (USP<1911>).
[5] More solutions to sticky problems: A guide to getting more from your Brookfield viscometer. Application note. Brookfield Engineering Laboratories Inc., MA, USA.
[6] Oscillation measurements in controlled stress rheology. Application note. DE, USA: TA Instruments Inc.
[7] Basic rheological terms. NJ, USA: Rheosys LLC, www.rheosys.com.
[8] Laba D. Rheological properties of cosmetics and toiletries, vol. 5. New York: Marcel Dekker Inc.; 1993.
[9] Profiling V, Shear Z, Curves F. Temperature, thixotropy & oscillation rheology & viscosity & profiling techniques & methods. Centre of Industrial Rheology, www.rheologyschool.com.
[10] Yield Stress Calculation Methods, www.azom.com.
[11] Product literature. DE, USA: TA Instruments Inc.
[12] Race S. Finding the right test protocol for polymer QC. Am Lab June 2002:26–7.
[13] Rheological analysis of dispersions by frequency sweep testing using equipment from Malvern Instruments, www.azom.com.
[14] Mazzeo FA. Importance of oscillatory time sweeps in rheology. Application note RH081a. TA Instruments Inc., DE, USA.
[15] Hu Y, Puwar K. Use of rheology in hot melt PSA formulation. NJ, USA: National Starch & Chemical Company (a subsidiary of Henkel AG & Co. KGaA).
[16] Rheological applications in pharmaceuticals and cosmetics. Bohlin Instruments Ltd; August 2001 (Version 2). (Reproduced under permission from Malvern Instruments Ltd. Malvern, Worcestershire, UK.)
[17] Rheology and Viscosity Testing Tips: Rheology of creams, ointments and lotions. Rheology of creams, ointments and lotions. Centre of Industrial Rheology, www.rheologyschool.com.
[18] Rheology oscillation and thixotropy. AN009, rev0, www.escubed.co.uk.
[19] Boger D, Scales P, Sofra F. Rheological concepts. Pastes and thickened tailings – a guide. 2nd ed. pp. 25–37.
[20] Jiao J, Burgess DJ. Rheology and stability of water-in-oil-in-water multiple emulsions containing Span 83 and Tween 80. AAPS PharmSci 2003;5(1):62–73. Article 7.
[21] Shon S-G, Lee C-H. A study on flow properties of semisolid dosage forms. Arch Pharm Res 1996;19(3):183–90.
[22] Yu L, Bunge A, Wilkin J. Bioequivalence of topical products. Presentation. www.fda.gov.

CHAPTER 10

Microscopy Techniques

Contents

10.1 INTRODUCTION

Microscopy is a very useful tool in pharmaceutical research, formulation development, process development, and quality control of manufactured products. Microscopy is routinely used for the examination of the drug material (the API—Active Pharmaceutical Ingredient), excipients (inactive ingredients present in the finished dosages), or the drug product [1,2]. Currently, there are two main types of microscopes classified based on the illumination sources—optical and electron microscopes. Optical microscopes use light sources in the visible range of the electromagnetic spectrum, whereas electron microscopes use an electron beam. Microscopes that use illumination sources in the ultraviolet (UV) and the infrared (IR) ranges are also available and their application in pharmaceutical product development is gaining popularity. Other, relatively newer, high–resolution microscopy techniques such as scanning tunneling microscopy and atomic force microscopy (AFM) are available, and have been used in the research phase of nanoparticle-related drug delivery systems. From the practical standpoint,

Essential Chemistry for Formulators of Semisolid and Liquid Dosages
http://dx.doi.org/10.1016/B978-0-12-801024-2.00010-8

183

optical microscopy is still the most widely used imaging technique in pharmaceutical product development and quality control.

10.2 OPTICAL MICROSCOPY

Optical microscopy is the most widely adopted imaging technique used in pharmaceutical product development and quality control. The reasons for this are the versatility of application for raw materials and finished products in their native state, ease of use, time taken to generate data, and cost. One of the primary applications of optical microscopy is to characterize particles of the raw material either by itself or in the finished product. This characterization includes generating information relating to the size of particles, particle morphology, and crystallinity.

10.2.1 Particle Size Testing

Raw materials, as well as suspended particles in the formulation (e.g., gels and ointments), can be characterized by optical microscopy. The resolution, d, of an optical microscope is given by the equation $d = 0.61 * \lambda / NA$, where λ is the wavelength of light in μm and NA is the numerical aperture (μm). In practical terms, the smallest particles that can be detected by optical microscopy are approximately 1 μm.

There are several software packages available that give an automatic count and determine the size distribution and morphology of particles. However, all of them have limitations when it comes to identifying particle boundaries, and distinguishing between the particle of interest and other particles. As a result, one critical aspect of measuring particle size by microscopy is the preparation of the sample slide. If the sample to be tested consists of well-separated particles, the software can count thousands of particles within a minute and provide meaningful size distribution data. However, the preparation of slides in which the particles are well separated becomes challenging when testing gels, ointments, or products with a viscous matrix. Consequently, fully automated particle counting and characterization may not be suitable for all types of samples because the data generated may not be accurate. To address this situation, software packages generally provide an option to manually correct the data. This process, however, can become very time-consuming due to the large number of particles that need to be measured to generate accurate/representative values for size. As for determining the number of particles to count, if all the particles are of the same size, just one particle is enough. In reality, however, this is not the case. Statistical software packages can be used to calculate the number of particles (sample size) required based on the accuracy to which the particle size is needed (e.g., the required standard error of the mean, standard deviation, and desired confidence interval). International Organization for Standardization (ISO) 13322-1 provides guidance on the number of particles required to be counted based on a geometric standard deviation and a 95% confidence level. However, products are generally mixtures of both coarse and fine particles; therefore, as part of method development, one may establish an optimum number of particles to be counted by comparing the results obtained from counting different numbers of particles—for example, in the range of 100–5000 particles. When the size distribution is independent of the number of particles counted, the data are likely to indicate a representative distribution of the material in question.

Although U.S. Pharmacopeial Convention (USP) ⟨776⟩ provides some guidance on particle size limit tests, the description of particle shapes, and the measurement of irregular particles, it does not provide guidelines regarding method validation for particle size measurement, or the acceptable variance. For quality control testing under Current Good Manufacturing Practice (CGMP) conditions, the method for measuring particle size must be validated. Accuracy, sensitivity, specificity, robustness, and reproducibility need to be established as part of method development and validation. Under USP ⟨1225⟩ analytical method classification, particle size is a Category-III method, for which establishing precision is critical for validation. National Institute of Standards and Technology traceable standard particles can be used to verify the calibration and normal performance of the microscope and software. A placebo sample can be spiked with standard particles to establish accuracy and sensitivity. In the absence of USP guidelines for acceptable differences between measured and certified values for these standard particles, self-established (and justified) acceptable variance can be used.

10.2.2 Morphology

The quantitative evaluation of particle shape is critical when powder flow is key to the performance of the product, such as in the case of dry powder inhalation dosage forms. Some common particle descriptors from USP ⟨776⟩ are:

- Lamellar: Stacked particles.
- Aggregates: Mass of adhered particles.
- Agglomerate: Fused or cemented particles.

- Conglomerate: Mixture of two or more types of particles.
- Spherulite: Radial clusters.
- Drusy: Particles covered with tiny particles.
  Particle morphology is quantitatively evaluated using various geometrical factors. ISO 9276-6 provides descriptive and quantitative representations of particle shape and morphology.
- Feret diameter: The distance between two parallel tangents on opposite sides of the image of a randomly oriented particle. The maximum Feret's diameter, F_{max}, also called the maximum distance in some references, is defined as the longest distance between any two parallel tangents on the particle. Likewise, the minimum Feret's diameter, F_{min}, also called the minimum distance in some references, is defined as the shortest distance between any two parallel tangents on the particle.
- Martin diameter: The diameter of the particle at the point that divides a randomly oriented particle into two equal projected areas.
- Length: The longest dimension from edge to edge of a particle oriented parallel to the ocular scale.
- Width: The longest dimension of the particle measured at right angles to the length.
- Convex hull perimeter: Length of an imaginary rubber band stretched around the particle.
- Convex hull area: Area covered within the convex hull perimeter.
  Commonly used nomenclature for shape descriptions are:
- Acicular: Slender, needle-like particle of similar width and thickness.
- Columnar: A long thin particle with width and thickness greater than those of an acicular particle.
- Flake: Thin flat particle.
- Plate: Flat particle of similar length and width, however, thicker than a flake.
- Lath: Long thin blade-like particle.
- Equant: Cubical or spherical particle.
  Commonly measured shape factors are [3,4] circularity (roundness), convexity, elongation, aspect ratio, diameter of circle of equivalent area, sphericity, and ratio of diameters of inscribed and circumscribed circles. Different shape factors for characterizing particles by image analysis are shown below.
- Aspect ratio $= F_{Min}$ length$/F_{Max}$ length
- Elongation $= 1-$(Aspect ratio)
- Sphericity $= 4\pi A/P^2$ (in which A = area and P = perimeter of the particle)

- Circularity is defined as the degree to which the particle is similar to a circle. Circularity $= P_c/P$ (in which P_c = perimeter of a circle of equal area as the particle, P = perimeter of the particle). Also Circularity $= \sqrt{}$(sphericity)
- Roundness $= 4A/\pi(L_{max})^2$ (in which A = particle area, $L_{max} = F_{max}$ length)
- Convexity, C_x, is a measurement of the particle edge roughness. $C_x = P_h/P$ in which P_h is the convex hull perimeter and P is the actual perimeter
- Solidity (compactness), S, is a measure of overall concavity of the particle. $S = A/A_c$, in which A = area of the particle and A_c = Convex hull area.

Figure 10.1 shows the convex hull. Table 10.1 shows shape factors for some of the common shapes.

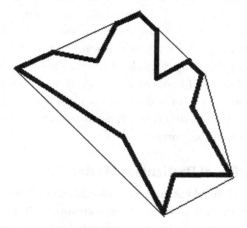

Figure 10.1 An imaginary particle is shown by the bold boundaries. The convex hull perimeter is shown as the line joining corners to corners, as if an elastic rubber band is stretched around the particle projection.

Table 10.1 Shape factors for some common geometrical shapes and objects

Shapes	Aspect ratio	Solidity	Circularity	Convexity
Circle	1	1	1	1
Square	1	1	0.89	1
Hexagon	0.9	1	0.95	1
Ribbon	0.11	0.85	0.49	0.99
Fiber	0.2	0.1	0.14	0.89
Equilateral triangle	1	1	0.78	1
A star-like object	0.81	0.69	0.49	0.63

10.2.3 Thermal or Hot-Stage Microscopy

Controlled heating and cooling stages are available for optical microscopes. These accessories can be used to heat specimens up to 350 °C in a controlled manner. Temperatures can be held fixed for a length of time to allow any transitions occurring within the specimen (active drug or excipient) to be observed. Hot-stage microscopy is a very useful tool during product development as well as for characterizing the finished product for thermal behavior. For example, with semisolid dosage forms, the drug substance might dissolve in the excipients at high temperature. When the product is subsequently cooled to room temperature, the drug may crystallize out. The crystalline form of the recrystallized drug, however, may be different from the initial form. In this situation, preformulation studies by hot-stage microscopy can provide critical information relating to the recrystallized drug substance or excipient. For ointments containing a suspended drug substance, particle imaging by hot-stage microscopy can be performed at elevated temperatures (40–50 °C). At this temperature the ointment matrix becomes fluid and the drug particles become clearly visible through the microscope. A calibrated hot-stage accessory is also very useful for determining the melting point of a drug or an excipient, and can be used to distinguish between two different types of suspended particles provided they have distinct melting points.

10.2.4 Evaluation of Particulate Matter

USP ⟨788⟩ (Particulate Matter in Injections) Method 2 is a microscopic particle count test. A known amount of test solution is filtered through a 1 μm (or finer) filter membrane, and is then examined using an optical microscope. Particles of 10 μm or larger, and 25 μm or larger, are counted. The product complies with the test if the average number of particles present in the units tested do not exceed 3000 per container that are equal to or greater than 10 μm, and 300 per container that are equal to or greater than 25 μm.

10.2.5 Evaluation of Foreign Particulate Matter

The optical microscope will probably be the first tool used in any investigation of the presence of "foreign" particulate matter found in development or commercial products. Optical microscopy under reflectance mode can provide information relating to the surface features of the particles, and may help in assessing the nature of the subject particle (e.g., metallic or nonmetallic, crystalline/birefringent or amorphous).

10.2.6 Evaluation of Material Compatibility

The physical compatibility of manufacturing components (e.g., mixing tank scraper blades, gaskets, seals, transfer pipes) or the container closure system with the formulation may be evaluated by microscopic examination. A sample of the component in question is placed in contact with the formulation for a given period of time and then examined for any discoloration, deposits, or defects compared to the "control" material.

10.2.7 Testing of Tablets

The US FDA has released guidance [5] relating to the size and shape of generic tablets and capsules. In this testing, optical microscopy will be a critical tool for the quantitative assessment of the size and shape parameters.

10.2.8 Ingredient-Specific Particle Sizing

Raman chemical imaging coupled to optical microscopy can be used to determine the chemical identity of different particles found during microscopic evaluation (e.g., in cases in which drug material is mixed with excipient particles). Use of this coupled technique also enables the particles of interest to be selected for size analysis [6]. Although this technique appears to be very appealing, it is not necessarily applicable to all sample types (e.g., where particles are embedded in a complex matrix such as a gel or ointment).

10.2.9 Fourier-Transform Infrared Microscopy

Fourier-Transform Infrared (FTIR) microscopy can be used to help detect any defects in a specimen—for example, in pharmaceutical package materials, or contamination in a formulation [7]. With this technique, a small area of the specimen is selected and an FTIR spectrum is recorded. The FTIR microscopy technique is also suitable for characterizing drug materials for polymorphism and hydration content.

10.3 SCANNING ELECTRON MICROSCOPY

In scanning electron microscopy (SEM), a focused beam of high-energy electrons (e.g., 20 keV) is scanned across the surface of the sample [7]. When these high-energy electrons strike the sample, three types of signal are generated—secondary electrons, backscattered electrons, and X-rays. These signals reveal details regarding the morphology and elemental composition of the sample.

Secondary electrons are emitted from the atoms occupying the top surface of the sample. Backscattered electrons are primary beam electrons which are "reflected" by atoms in the solid. Both secondary and backscattered electrons are used to produce a readily interpretable image of the surface. The image produced is of very high magnification (100,000×) with a resolution of approximately 5 nm.

Interaction of the primary beam with atoms in the sample causes excitation of the orbital electrons. When these excited electrons return to a lower energy state they emit X-rays of fixed wavelengths. The emitted X-rays have an energy that is characteristic of the parent element. Detection and measurement of this energy permit elemental analysis (Energy Dispersive X-ray Spectroscopy [EDX or EDS]). SEM analysis is "non-destructive"—that is, there is no sample loss and the sample may be analyzed multiple times, if needed.

SEM is commonly used in product development to analyze the morphology and size of nanoparticles. SEM/EDX is a very useful tool for investigating any "foreign" particulates. The limitation of SEM is that the sample needs to be placed under high vacuum; therefore, this technique is not suitable for wet samples.

10.4 TRANSMISSION ELECTRON MICROSCOPY

Transmission electron microscopy (TEM) is a very powerful research tool in which a highly accelerated electron beam (100–1000 keV) is passed through an ultra-thin sample to reveal the details of nanostructures (such as liposomes, solid lipid nanoparticles). TEM requires extensive skills with respect to both sample preparation and operation of the instrument. Sample preparation involves depositing a thin film of carbon on the TEM grid, placing the liquid sample (e.g., liposome solution) on the carbon film, and then blotting away the excess liquid. The background of the sample is then stained with a staining agent such as uranyl acetate solution, forming a "shadow" around the sample (this is known as negative staining). The sample appears as a negative picture when the grid is examined under a microscope.

To visualize the nanostructures in their native state (i.e., without staining), a technique known as Cryo-TEM is used. In Cryo-TEM, the sample is mounted on a perforated grid and examined at a very low temperature (~100 K). The analyst looks for the nanostructures of interest overlying one of the holes in the supporting carbon film.

Structural details of the internal and/or external parts of the nanoparticles can be obtained using Freeze–Fracture TEM (FF-TEM). The solution

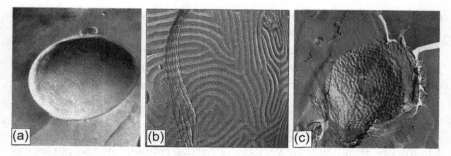

Figure 10.2 Freeze-fracture micrographs of liposomes showing surface texture at room temperature. (a) Liposome of palmitoyl oleoyl phosphatidylcholine (POPC) showing a smooth surface texture. (b) Liposome of a mixture of POPC and Galactosylceramide showing surface undulations in which the undulations are registered through several layers of the liposome. (c) Liposome of 24:1 Sphingomyelin showing a rough surface texture.

of nanoparticles is frozen in liquid ethane, and then held and fractured at liquid nitrogen temperature. The fractured sample is coated at an angle with carbon, and the carbon film (which is a replica of the sample) is then transferred to an electron microscope grid and examined by TEM at room temperature. In this method, it is the replica of the sample that is examined and not the sample itself. FF–TEM is very useful for viewing the internal core of liposomes or nanoparticles. Figure 10.2 compares the undulated surface texture of the vesicles formed by (1) a mixed lipid system of palmitoyl oleoyl phosphatidylcholine (POPC) and galactosylceramide (GalCer), and (2) pure 24:1 sphingomyelin to the smooth surface exhibited by POPC alone.

The limitations of the various TEM techniques are that they require a very high level of operator training and skill, and are very time-consuming and expensive.

10.5 ATOMIC FORCE MICROSCOPY

Scanning probe microscopy covers a group of technologies that are used for imaging and measuring surfaces at a molecular/group of atoms level. In AFM, the surface of the specimen is scanned with an extremely fine tip (often less than 100 Å in diameter and approximately 2 μm in length) that is mounted on a flexible cantilever, thus allowing the tip to follow the surface profile of the sample. Van der Waals forces between the tip and the surface of the specimen cause the cantilever to bend or deflect. A sensor measures the deflection of the cantilever as either the tip is scanned over the sample, or the sample is scanned under the tip. Based on the movement of the cantilever, a topographic map (image) is generated. AFM can be operated in a "contact" or "non-contact" mode.

The use of AFM in the pharmaceutical field is not widespread in either product development or manufacturing. Nonetheless, it is a very useful tool during the research phase for characterizing nanodrug delivery systems [8,9].

10.6 CONCLUSION

Microscopy techniques are very useful for determining the morphology of particles. The advantages of using microscopy for particle size analysis are that it can be ingredient specific, and is complementary to other particle sizing techniques. As such, microscopy may be the most applicable technique for determining the size of particles within some semisolid products, such as gels and ointments. The challenges of using microscopy for particle sizing are (1) deciding the sample size (number of particles) for accurate size analysis of the product, and (2) the level of skill required by the operator.

REFERENCES

[1] Ku MS, Liang JQ, Lu D. Application of microscopy in pharmaceutical development from discovery to manufacture process scale-up. Microsc Microanal July 2010;16(Suppl. S2):636–7.
[2] Carlton RA. Pharmaceutical microscopy. Springer; 2011.
[3] Houghton ME, Amidon GE. Microscopic characterization of particle size and shape: an inexpensive and versatile method. Pharm Res 1992;9(7):856–9.
[4] Olson E. Particle shape factors and their use in image analysis – Part-1: theory. J GXP Compliance 2011;15(3):85–96.
[5] FDA Guidance for Industry. Size, shape, and other physical attributes of generic tablets and capsules. June 2015. Available at: www.fda.gov.
[6] Doub WH, Adams WP, Spencer JA, Buhse LF, Nelson MP, Treado PJ. Raman chemical imaging for ingredient-specific particle size characterization of aqueous suspension nasal spray formulations: a progress report. Pharm Res 2007;24(5):934–45.
[7] Andria SE, Fulcher M, Witkowski MR, Platek SF. The use of SEM/EDS and FT-IR analyses in the identification of counterfeit pharmaceutical packaging. Am Pharm Rev April 2012;15(3).
[8] Grobelny J, DelRio FW, Pradeep N, Kim DI, Hackley VA, Cook RF. Size measurement of nanoparticles using atomic force microscopy. Methods Mol Biol 2011;697:71–82.
[9] Van Eerdenbrugh B, Lo M, Kjoller K, Marcott C, Taylor LS. Nanoscale mid-infrared imaging of phase separation in a drug–polymer blend. J Pharm Sci 2012;101:2066–73.

CHAPTER 11

Miscellaneous Physical, Chemical, and Microbiological Test Methods

Contents

Essential Chemistry for Formulators of Semisolid and Liquid Dosages
http://dx.doi.org/10.1016/B978-0-12-801024-2.00011-X

This chapter covers test methods typically used to evaluate and characterize semisolid formulations. The methods have been subdivided into physical, chemical, and microbiological, as well as separating them into tests that might typically be performed on all semisolid formulations, and those that are specific to a type of product. When they are performed during the product lifetime, and what information they can provide, is also described.

11.1 GENERAL PHYSICAL TESTS

This section covers physical tests that might be performed on any semisolid formulation, and are summarized in Table 11.1.

11.1.1 Description/Appearance

This is usually a general statement describing how the product should look when visually examined—for example: for a gel "Clear colorless to straw colored gel, free from particulate matter"; for a solution "Clear, colorless solution essentially free from particles of foreign matter"; for a suspension "White to straw-colored suspension free of foreign particles"; for an ointment "A white to off-white ointment, smooth and homogeneous in appearance, odorless or with a slight petrolatum odor." Any changes to the visual appearance of the product would indicate gross problems during processing or on stability.

Table 11.1 General tests—physical

Test	Comments	Development tests	Bulk	Finished product	Stability
Description/appearance			X	X	X
Color	Aqueous formulations only		X	X	X
pH			X	X	X
Osmolality			X	X	X
Viscosity			X	X	X
Rheology		X	X	X	X
Specific gravity					
Net content/minimum fill				X	
Particle size distribution			X	X	X
Particulate matter			X	X	X
Weight loss			X	X	X
Container closure integrity			X	X	X
Temperature cycling		X	X	X	
Freeze/thaw		X	X	X	
Photostability		X		X	
Conductivity				X	X
Microscopy				X	X
In vitro release testing		X	X	X	
Transit trials		X		X	
Differential scanning calorimetry/thermal analysis		X	X	X	X

11.1.2 Color

Color testing could be a visual assessment, or a quantified measurement using a colorimeter (e.g., a Lovibond® [1]). With pharmaceutical and cosmetic semisolid formulations, color is an important means of determining the identification and aesthetic value of the product. In certain circumstances, it can also be used as a way to measure the progress of processing. Changes in color, either during processing or on storage, are an indicator of degradation of the formulation. Grading techniques are widely used to assess product color by comparison with a representative series of fixed color standards.

11.1.3 pH (US Pharmacopeia [USP]<791>)

pH measurements are typically made using a calibrated pH meter at 25 °C (or other specified temperature), and give a measure of the acidity or basicity of an aqueous solution. The pH stability profile of the active ingredient within the formulation, combined with the safe pH range for the product in use, will define the acceptable limits for the formulation. pH measurements can be taken on the bulk product (to assess the extent of any adjustments to be made prior to completion of mixing and commencement of filling), on the finished product (to determine compliance with set limits, and release of the batch), and on storage (to give an indication of formulation stability).

11.1.4 Osmolality (USP<785>)

Osmolarity and osmolality are units of solute concentration that are often used in reference to biochemistry and body fluids, and are related to the tonicity of the formulation. If the tonicity is too far from isotonic, certain products (e.g., ophthalmic solutions and suspensions) will cause stinging on application. Sodium chloride is often used to adjust the osmolality of a formulation. Osmolality is measured using an osmometer by, for example, freezing point depression of the solution. As with pH, osmolality can be measured and adjusted on the bulk material, and measured and monitored on the finished product at the time of release and on storage.

11.1.5 Viscosity (USP<911>, <912>, <913>)

Viscosity testing is discussed at length in Chapter 9. Viscosity testing is typically used as a routine Quality Control test on bulk material, on the finished product at the time of release (to monitor any batch-to-batch changes in the formulation) and on storage (to give an indication of formulation stability).

11.1.6 Rheology (USP<912>)

Rheology testing is also discussed at length in Chapter 9. Rheology testing is typically used during formulation development—the information generated being used to compare different formulations, compare the effect of different formulation ingredients (e.g., thickeners), evaluate the effect of the manufacturing process, and evaluate formulation stability.

11.1.7 Specific Gravity (USP<841>)

Many semisolid formulations are sold (to the end user) by volume. However, most in-process fill measurements are by weight. The specific gravity of the bulk formulation is determined to enable this conversion. The specific gravity of, for example, gels and emulsions is also used to check that excessive amounts of air (bubbles) have not been incorporated into the formulation during processing—excess air can lead to spoiling and degradation of ingredients, and affect other physical measurements (e.g., viscosity).

11.1.8 Net Content/Minimum Fill (USP<755>)

The USP [2] specifies requirements for minimum fill testing for products such as creams, gels, jellies, lotions, ointments, pastes, powders, and aerosols. This testing is performed on packaged product, either as an in-process control or on batch release.

11.1.9 Particle Size Distribution

The size distribution of suspended particles in semisolid formulations (gels, suspensions, and emulsions) can be important to the efficacy and stability of the product. For a given concentration of a material, smaller particles have a larger surface area than larger particles. As the "activity" of a material is related to its surface area, it may be important to characterize and quantify this parameter. During processing, steps can be incorporated to reduce the particle size of suspended particles to the acceptable range. In-process testing of bulk material can serve to follow the particle size reduction process, and finished product testing can be used to release the batch. Testing the particle size distribution on storage will provide stability information on any agglomeration, flocculation, etc. within the formulation.

Both manual (by light microscopy, as described in USP <776> and <788> method 2) and automated (by laser diffraction) techniques are available.

Particle size distribution testing is not applicable for ingredients that are in solution.

11.1.10 Particulate Matter (USP<788>, <789>)

Although USP<788> is written for injectable formulations and USP<789> for ophthalmic solutions, in the absence of specific guidelines, they have been adopted as a basis for other types of formulations—for example, nasal sprays. The test procedures and equipment are detailed in USP<788>, the permitted limits being covered in the individual chapters.

Particulate matter is defined as being extraneous to the formulation—that is, not part of the formulation, not a suspended ingredient. Particulate matter testing is typically carried out on the finished packaged product, such that contributions from all of the manufacturing processes and the packaging itself are taken into account. Any increase in the number of particles detected on storage would indicate either that ingredients were coming out of solution (crystallizing), or that the packaging was breaking down and starting to shed particles into the product.

11.1.11 Weight Loss

Weight loss testing is carried out on storage of a product, to assess whether water (or other volatile ingredients) is being lost from the formulation, through the packaging. If loss is occurring, the formulation will change with time. This will lead to a concentrating effect on the remaining ingredients, and may cause changes such as an increase in viscosity/hardening of the formulation, loss of solubility/crystallization of ingredients, or even the formulation drying out.

Weight loss testing is nondestructive, and is performed by identifying a number of units at time zero, recording their starting weight, and then reweighing the same units over the shelf life of the product.

11.1.12 Container Closure Integrity

Many methods are available for testing container closure (or packaging) integrity. Some of these methods have been summarized below. The intent of container closure integrity testing is to ensure that the container will adequately protect the product, and that the integrity will not degrade over the lifetime of the product.

- Dye ingress test: involves immersing the package in dye solution (e.g., methylene blue) in a vacuum chamber. A specified vacuum is pulled and held for a given period of time. After the vacuum is released the test articles are removed, cut open (if not transparent), and inspected. No dye should be seen inside the container. Positive controls, that have been compromised (e.g., by piercing with a needle), should be run at the same time. The positive controls should show the presence of dye inside the packaging. This type of test can be used for any packaging that will not be destroyed by immersing in liquid.

- Vacuum test: involves immersing the package in water in a vacuum chamber. A specified vacuum is pulled. A stream of bubbles escaping from the package indicates a compromised pack. This type of test can be used for any packaging that will not be destroyed by immersing in water.

- Squeeze test: involves applying a fixed external pressure to the container. Any signs of product leaking from the container, or decay of applied pressure, indicate a compromised pack. This type of test can be used for packaging that is intended to be squeezed to dispense the product—for example, plastic tubes with a heat seal, sachets, plastic bottles. This type of test is not appropriate for metal tubes with a crimp seal because the applied pressure will cause the crimp to unwind—which constitutes a different type of failure.

- Vacuum decay (using external vacuum): is an automated noninvasive and nondestructive test. The package is placed inside a test chamber, and a vacuum is applied. A vacuum transducer is used to monitor the test chamber for both the level of vacuum as well as the change in vacuum over a predetermined test time. Changes in absolute and differential vacuum indicate the presence of leaks and defects within the package. This type of test can be used for most types of rigid, semirigid, and flexible packages. Several different makes and models of equipment are available—including VeriPac® [3], Uson Sprint® iQ and Uson Optima® leak testers [4], and Qualipak® [4] package chambers.

- Pressure decay (using introduced pressure): is a semiautomated destructive test. A preselected pressure is introduced into the package (e.g., using a needle to puncture the pack) and then held at a constant level for the test time. Any pressure loss during the test is an indication of a compromised pack. This type of test can be also be used for most types of rigid, semirigid, and flexible packages. Examples of this type of equipment are the Lippke® Model 4000 and Lippke® Model 4500 leak detectors and seal strength analyzers [5].

11.1.13 Temperature Cycling and Freeze/Thaw

As mentioned in Section 7.4.5.1, temperature can have a major effect on the chemical and physical stability of a formulation. Chemically, a 10 °C rise in temperature leads to an approximate doubling in the rate of reaction. Physically, increasing the storage temperature typically reduces the viscosity of a semisolid, which can accelerate the migration of particles and droplets leading to sedimentation or phase separation (particularly for gels, ointments, and emulsions). Cycling the temperature between ambient room temperature and 40 °C can exaggerate physical effects, providing results in a shorter timescale.

Freeze–thaw cycling can also destabilize aqueous gels and emulsions by the formation of ice crystals—leading to "cracking" and phase separation within the formulation.

Temperature cycling and freeze/thaw studies are typically performed during formulation development—to give an indication of the stability of different formulation candidates. For this testing, the bulk formulation can either be stored in glass jars (to permit the visual inspection of the formulation without unduly disturbing it), or in the final packaging (if available).

11.1.14 Photostability

Photostability testing is used to determine the susceptibility of a formulation to photodegradation by accelerating possible chemical instabilities/interactions (see Section 7.4.5.1 for test conditions). As with temperature cycling and freeze/thaw studies, photostability testing is typically performed on the bulk formulation stored in glass jars during formulation development (so as not to impede the exposure to light). The information generated by this analysis can then be used during the design of the packaging—for example, if the formulation is light sensitive, opaque or light-proof containers can be used. Once the packaging is finalized, the photostability exercise is repeated to assess the effectiveness of the pack design.

11.1.15 Conductivity

Conductivity can be used as a way of determining the gross uniformity of a semisolid formulation (gel or emulsion) in either the bulk mixing tank, or in the packaging. The conductivity is measured at the top and bottom of the sample at each time interval (e.g., during the mixing process, or on stability storage), and any differences in the conductivity will indicate a nonhomogeneous sample. This technique can also be used during bulk hold studies of the formulation, prior to packaging.

11.1.16 Microscopy

Evaluating semisolid formulations using a microscope can give information relating to the efficiency of processing and the stability of suspensions, gels, emulsions, and ointments. It can also be used to help identify particulate contaminants. This testing can be performed at any stage of the manufacturing process or product lifetime—for example, to investigate:

- Mixing of emulsions—size, homogeneity, and uniformity of distribution of internal phase droplets
- Dispersion of suspended particles—homogeneity of particles
- Particle size reduction—size and uniformity of particles
- Particulate matter investigations—crystal shape/morphology, birefringence (using cross-polarizers), and particle melting point (using a hot stage). Automated equipment is also available that will not only determine the size and shape of particles, but will identify selected particles by Raman spectroscopy (e.g., Morphologi G3® [6], rap.ID® [7])
- Formulation stability—changes in droplet and particle size and/or distribution with time.

11.1.17 *In Vitro* Release Testing

This test uses Franz Diffusion Cells to measure the rate of permeation of a formulation ingredient through a membrane (e.g., permeation of an active pharmaceutical ingredient through the skin) [8]. The test can be used to compare permeation rates of different formulations, or to compare generic formulations to the reference listed drug. Due to the difficulty in obtaining human skin, and safety concerns for the analyst using it, polymeric membranes are often used as a substitute.

Changes to the formulation, ingredient concentration, and adding permeability enhancers can affect the diffusion characteristics of the material of interest. This testing is typically performed during formulation development.

11.1.18 Transit Trials

Transit testing, simulated or real life, is performed during the product development process to assess the adequacy and durability of the packaging, to prevent unacceptable damage to the product. Protection of the product is important at all stages in the supply chain—from the point of manufacture, to "delivery" to the end user. Several test regimes (protocols) are available depending on the package and distribution route (e.g., package weight, whether shipped internationally, whether temperature control is needed) [9]. Consequently, it is important to understand the flow of packages—how they are shipped, handled, and stored (in the manufacturer's facility and in warehouses, transportation facilities, vehicles, and customer locations). It is also important to know the allowable level of damage to the product and packaging (e.g., minor dents to tubes).

Measurement instruments and techniques can be used to determine vibration, drop, compression, and temperature/humidity levels encountered during real life distribution. This information can then be used to generate test protocols in the laboratory. In the absence of this information, standardized test protocols are available to simulate transit conditions. The testing typically includes: drop and impact testing (to simulate accidents during handling); vibration testing (to simulate transportation); compression load testing (both static—to simulate stacking in a warehouse, and dynamic—under vibration, to simulate pallet stacking during transport); and a variety of different atmospheric conditions (e.g., temperature, humidity, and pressure).

In addition to performing transit testing during product development, any changes to the product or packaging during the product lifetime also need to be evaluated—for example: changes in the design, size/weight, or materials of construction (grade of board).

Examples of transit tests are given below:

- Vibration testing (using a vibration table): the frequency of vibration is adjusted until the test shipper repeatedly leaves the platform and a 1/16 inch thick × 2 inch wide shim can be inserted at least 4 inches between the bottom of the package and the surface of the table and intermittently moved along one entire length of the shipper. Vibrate the test shippers for 30 min, then rotate the test shippers 90° and vibrate for an additional 30 min. Visually inspect the shippers and the contents for any damage or deterioration resulting from the test.
- Drop testing: filled cases are dropped based on the case weight

Case weight	Free fall drop height
1 pound through 20.99 pounds	30 inches
21 pounds through 40.99 pounds	24 inches
41 pounds through 60.99 pounds	18 inches
61 pounds through 100 pounds	12 inches

Each test case is dropped 10 times as specified in the following sequence:
- Bottom corner at the case manufacture joint
- Shortest edge radiating from the corner drop point
- Next longest edge radiating from the corner drop point
- Longest edge radiating from the corner drop point
- Flat on one of the smallest faces
- Flat on the opposite small face

- Flat on one of the medium faces
- Flat on the opposite medium face
- Flat on one of the largest faces
- Flat on the opposite largest face.

After all drops have been completed, the shipper and its contents are visually inspected for any damage or deterioration resulting from the test.

11.1.19 Differential Scanning Calorimetry/Thermal Analysis (USP<891>)

As a sample is heated, thermodynamic events such as a change of state, are accompanied by a change in energy within the system. These exothermic or endothermic transitions can be observed using techniques such as differential scanning calorimetry and differential thermal analysis. These techniques measure temperatures and heat flows associated with thermal transitions in a material. Common uses include investigation, selection, comparison, and end-use performance evaluation of materials in research, quality control, and production applications [10]. Another thermoanalytical technique, thermogravimetric analysis, measures weight changes in a material as a function of temperature (or time) under a controlled atmosphere. Its principal uses include measurement of a material's thermal stability and composition.

The various thermoanalytical techniques provide information on desolvation, dehydration, decomposition, crystallization, polymorphism, melting temperature, sublimation, glass transitions, evaporation, pyrolysis, solid–solid interactions, phase changes, and purity of ingredients. These techniques can be used at any stage during the product life cycle—from formulation development (to provide insight into formulation structure and changes during processing, and to compare between different formulations and ingredients), to routine quality control (to assess any batch-to-batch differences) and commercial stability testing.

11.2 GENERAL CHEMICAL TESTS

This section covers chemical tests that might be performed on any semisolid formulation, and are summarized in Table 11.2.

11.2.1 Identification Test

This test is performed to positively identify the presence of the active ingredient(s) in a pharmaceutical formulation—that is, to ensure that the correct pharmaceutically active ingredient(s) has been used during manufacture.

Table 11.2 General tests—chemical

Test	Development tests	Bulk	Finished product	Stability
Identification		X	X	X
Active ingredient assay		X	X	X
Active ingredient degradants and impurities assay		X	X	X
Preservative assay		X	X	X
Other chemical assay (e.g., EDTA, ethanol)		X	X	X
Content uniformity			X	X
Extractables/leachables	X		X	X

Identification testing is performed at all stages of the product lifetime to prevent mislabeling/mishandling and to ensure user safety. If the product has a compendial monograph, the identification tests might be specified in the monograph. Typical identification tests are by IR scans, or high-performance liquid chromatography (HPLC) retention times (from the active ingredient assay).

11.2.2 Assays for Active Ingredient, Degradants, and Impurities

These assays are performed to ensure that the correct amount of active ingredient has been added during the manufacture of the formulation, and that it remains within acceptable limits throughout the shelf life of the product. Degradants are formed if the active ingredient degrades during processing or on storage—for example, due to sensitivity of the active ingredient to environmental conditions such as temperature, the presence of oxygen, or exposure to light. Degradants can also be formed from chemical reactions between the active ingredient and other formulation ingredients—for example, pH, or the presence of metal ions. These effects are usually identified and minimized during the formulation development process—by adjusting the pH to a more "active ingredient friendly" range, using a pH buffer system, adding a chelator, adding an antioxidant, and using light-proof packaging. Impurities, on the other hand, would be present in the active ingredient from its manufacturing process, and would be carried over into the final formulated product.

Active ingredient, degradants, and impurities assays are performed at all stages of the product lifetime:

- on bulk product—to assess the homogeneity of the batch/mixing efficiency, and that the correct amount of active ingredient has been added.

As products progress through the different manufacturing stages (from bulk formulation, to packaged product in the factory, to product out in the marketplace), more costs are incurred. Testing product at this stage, and formally releasing the product to further processing, reduces the cost risk. Reworking or reprocessing might be possible at this stage of the process. However, once the formulation is packaged, the possibility of reworking becomes less

- on finished product—to demonstrate that the formulation has not been adversely affected during the packaging operation, that the batch is homogeneous throughout the entire filling process, and that the batch complies with approved specifications (e.g., no sedimentation during storage, confirmation of amount of active ingredient added). This testing would constitute part of the release testing of the product, and would include testing packages throughout the packaging run (e.g., beginning, middle, and end)
- on stability—to monitor the buildup of any degradants over the shelf life of the product. Degradation products might have a different toxicity profile compared to the active ingredient, hence the use of a stability indicating assay will provide data to show whether the product will remain safe and effective over time.

Typically HPLC or UPLC methods are used for assaying for the active ingredient, degradants, and impurities. The stability indicating assay used to determine degradants, however, might be different compared to the active ingredient assay.

11.2.3 Preservative and Other Chemical Assays

The level of preservatives and other functional ingredients (such as ethanol, chelator, antioxidant) can also be tested at all stages of the product lifetime. These ingredients are included to protect the product from degradation caused by microbial contamination, metal ions, atmospheric oxygen, etc. The minimum level of these ingredients will be determined during formulation development, and will depend on the susceptibility of the formulation to the various methods of degradation combined with the method of use/application of the product. For instance, when a multidose nasal spray is sprayed from its container, air is sucked back into the container to replace the expelled product. The air that is sucked back into the device (unless passed through a sterilizing filter) will contain atmospheric microorganisms that have the potential to cause the product to spoil. This air will also contain oxygen, which could cause oxidation of the formulation. A unit-dose nasal spray does not experience this level of insult, and consequently does not need as robust a preservative or antioxidant system. On a similar note, a multiuse pressurized aerosol will prevent the ingress of oxygen by the fact that the pressure inside the container is higher than atmospheric pressure—so oxygen ingress will be prevented.

The reasons for testing the levels of these functional ingredients at the different stages of the product lifetime are similar to the case for the active ingredient.

11.2.4 Content Uniformity

Content uniformity testing is performed to show that the active ingredient, preservatives, and other functional ingredients are uniformly distributed within each individual container, as well as throughout the entire batch of packaged product. This is needed to ensure the quality, safety, and efficacy of the product—that is, that each dose of a product will be the same as the previous one, and that some containers from a batch are not superpotent whereas others are subpotent.

For example, in the case of a product in a tube (ointment, gel, cream), the container is cut open in the shape of an "I." Samples are then taken from the cap, middle, and crimp areas, and assayed. Any differences within the tube might indicate separation of the bulk formulation, or sedimentation of solid ingredients within the tube on standing.

As mentioned in Section 11.2.2, assaying containers throughout the entire packaging run (whether using a statistical sampling scheme, or testing beginning/middle/end of the operation) will provide information on the container-to-container product uniformity.

11.2.5 Extractables/Leachables

Extractable and leachable information should be generated during the product development stage to determine any interactions between the product formulation and the container closure system (packaging). This work would need to be reevaluated if any changes to the composition of the container closure system are made.

Extractables are compounds that can migrate from a material into a solvent under exaggerated conditions of time and temperature. Extractable data are, therefore, independent of the product. Leachables are compounds that actually do migrate into a product formulation under normal processing or storage conditions. All materials have extractables, but not all extractables will leach into the product [11]. Examples of materials that might be extracted from plastic container closure systems include plasticizers, lubricants, pigments, stabilizers, antioxidants, and binders. In the case of aluminum tubes, lacquers and coatings (e.g., epoxides) are used inside to prevent contact with the bare metal. Printing inks, varnishes, and adhesives (from the packaging artwork/labeling) also need to be assessed. These materials can migrate through plastic packaging on storage. Also, with some plastic tubes, the bulk roll-stock for the tube body is printed and then rerolled prior to the tube construction process. This rerolling brings the outer (printed) surface of the plastic into contact with the inner (product contact) surface, permitting transfer to occur.

During an extractable and leachable study, first the extractables are evaluated to screen for toxic materials. For example, a plastic component would be "extracted" using polar solvents, nonpolar solvents, and water under various conditions of heat and time. Any organic (semivolatile) materials extracted would be identified using high-performance liquid chromatography-photo diode array detection-mass spectrometry (HPLC-PDA-MS), organic (volatile) materials by gas chromatography-mass spectrometry (GC-MS), and metals by inductively coupled plasma-optical emission spectroscopy or mass spectrometry (ICP-OES or ICP-MS). Other techniques, such as ion chromatography and Fourier transform infrared spectrophotometry, might also be used to aid in the identification process. A toxicological assessment of the identified extractable materials would then be performed to determine which of these materials, if present in the product, would be of concern and at what level.

For the leachable study, product filled into the container closure system would be set up on stability (either real-time or accelerated), and then assayed for the presence and level of the chemicals of concern that were identified within the extractable work.

Numerous contract testing laboratories offer extractable and leachable testing services, including Pace Analytical Life Sciences, LLC. [12], Intertek [13], Eurofins Lancaster Laboratories [14], and SGS Life Science Services, a division of SGS North America Inc. [15].

Further information on extractable and leachable testing can be found in USP<661> (Containers-Plastic) and USP<381> (Elastomeric Closures for Injections).

11.3 GENERAL MICROBIOLOGICAL TESTS

This section covers microbiological tests that might be performed on any semisolid formulation, and are summarized in Table 11.3.

Table 11.3 General tests—microbiological

Test	Comments	Bulk	Finished product	Stability
Microbial limits testing		X	X	X
Antimicrobial effectiveness testing	Preserved products	X	X	X
Sterility	Sterile products		X	X
Endotoxins	Parenterals		X	X

11.3.1 Microbial-Limits Testing

The level of microbiological cleanliness specified for a product depends on the product in question and its intended use. Microbial-limits testing is performed for quality control purposes, to determine whether a formulation complies with established specifications. USP<61> (Microbiological Examination of Nonsterile Products: Microbial Enumeration Tests) covers the quantitative enumeration of mesophilic bacteria and fungi (yeasts and molds) that grow under aerobic conditions. If necessary, incubation under anaerobic conditions can also be performed. USP<62> (Microbiological Examination of Nonsterile Products: Tests for Specified Microorganisms) covers testing for the presence of specified microorganisms—"objectionable organisms."

For a nonsterile product, microbial-limits testing is performed throughout the lifetime of the formulation. Any microbial contamination of the formulation ingredients will be carried through to the final compounded bulk formulation. Testing the ingredients on incoming inspection will minimize this risk. Testing the bulk formulation will show whether microbial contamination has been introduced during the compounding process—either through the use of "dirty" equipment or the environment. Testing the packaged product will show whether the downstream processes and the packaging itself have added to the microbiological loading—and will also serve as one of the release tests for the product. Testing the product on storage will determine whether the microbiological quality of the product deteriorates over time—perhaps through compromised packaging.

In the case of sterile products, any sterilization process can only kill a finite number of organisms. Monitoring the bioburden will provide the quantity of viable microorganisms present before the sterilization process. Comparing this to the validated limits for the process will show whether it is within the acceptable range, or whether the sterilization efficiency will be compromised. Bioburden testing can also assist in monitoring the entire manufacturing process, from raw materials through clean room procedures.

11.3.2 Antimicrobial Effectiveness Testing

USP<51> reports that antimicrobial preservatives are added to nonsterile formulations "to protect them from microbiological growth or from microorganisms that are introduced inadvertently during or subsequent to the manufacturing process." In the case of sterile formulations that are packaged in multiple-dose containers, "antimicrobial preservatives are added to inhibit the growth of microorganisms that may be introduced from repeatedly withdrawing individual doses."

Antimicrobial effectiveness testing is performed at all stages of the product lifetime:

- during development: as mentioned in Section 7.4.6, batches of the candidate formulations are made at different preservative levels (e.g., 50, 75, and 100% of target) and tested for antimicrobial effectiveness. Making batches at a range of different preservative concentrations enables the minimum level needed to provide adequate protection to be determined. This value can then be used as a basis to set the minimum specification limit. Bulk product is typically used for this exercise
- bulk product: bulk product testing is typically only performed at the development stage, not during routine quality control
- finished product: antimicrobial effectiveness testing is performed on the finished product to ensure that the correct level of preservatives have been added during the manufacture of the formulation, and is used as a quality control test to release the product to the marketplace
- on storage: the assayed level of antimicrobial ingredients might reduce over the shelf life of the product—either due to the chemical degradation of the ingredient, or due to it being "consumed" while preserving the product (killing any microorganisms introduced from the other ingredients, process operations, or packaging components). This testing is performed over the shelf life of the product to ensure that, at the end of the shelf life, the product will still be adequately preserved—that is, above the minimum effective level determined during the development work

Antimicrobial effectiveness testing involves inoculating the formulation with given concentrations of specified microorganisms, incubating, and then determining the number of colony forming units (cfu) present in each test preparation at defined time intervals (using USP<61> and USP<62>). The results of the test are expressed as changes in $\log_{10}$ values of the concentration of cfu per mL for each microorganism at the applicable test intervals. Limits for the level of reduction are set in the compendial chapters.

11.3.3 Sterility Testing

Sterility testing is performed on sterile products (including ophthalmic and injectable formulations, as well as some topical preparations), and covers all types of semisolid and liquid formulations (USP<71>). Sterility testing is carried out after the sterilization process has been applied to the product as a finished product release test, as well as on storage over the shelf life of the product. A sterility test failure is evidence that either the sterilization process was not effective, or that the integrity of the packaging has been compromised. The test for sterility is carried out under aseptic conditions and involves inoculating the formulation into microbiological growth media, incubating, and assessing for any microbial growth (usually indicated by the growth media turning cloudy). The lack of any growth indicates that the product complies with the test for sterility.

11.3.4 Endotoxin Testing (USP<85>)

Endotoxins are found in the outer membrane of the cell wall of Gram-negative bacteria. They elicit a strong immune response in man (e.g., fever, septic shock), and cannot be removed from materials by normal sterilization processes. At the time of writing, endotoxin testing was a requirement for parenteral formulations, the limits for various products being listed in the individual pharmacopeial monographs. Eliminating Gram-negative bacteria from the formulation ingredients, packaging components, process equipment, and manufacturing environment will minimize their presence in the finished product. Any increase in the level of endotoxins on storage would indicate that either the product is not sterile and that the Gram-negative bacteria present are multiplying, or that the packaging has been compromised and allowed the influx of Gram-negative bacteria on storage.

Endotoxins can be measured using two methods [16]: (1) measuring a clotting reaction between the endotoxin and a clottable protein in the amoebocytes of *Limulus polyphemus* (the horseshoe crab); or (2) a more sensitive photometric test based on a Limulus amoebocyte lysate and a synthetic color-producing substrate.

11.4 PRODUCT-SPECIFIC TESTS

This section covers some of the tests that are performed on certain types of product, and are summarized in Table 11.4.

11.4.1 Tests Performed on Metered Dose Nasal Sprays and Pressurized Aerosols (e.g., pMDIs); Sublingual Sprays [17]

11.4.1.1 Spray Pattern

Spray pattern testing is performed to evaluate and characterize the performance of the spray pump. The spray pattern is the shape of the plume when looking directly towards the actuator orifice as the product is emitted from the device. Testing is performed at fixed distances from the actuator tip using a nonimpaction laser sheet-based instrument, and is characterized using the following parameters:

- D_{max}—the length of the longest line that passes through the weighted center of mass drawn within the perimeter of the spray pattern
- D_{min}—the length of the shortest line that passes through the weighted center of mass drawn within the perimeter of the spray pattern
- Ovality ratio—the ratio of D_{max}/D_{min}
- Ellipticity—the ratio of the major and minor axes of the ellipse

The size and shape of the actuator orifice, pump design, size of metering chamber (for multidose devices), and formulation characteristics (e.g., viscosity [18]) can affect the spray pattern.

This testing is performed on the finished product as a quality control release test, and on storage to determine any changes within the product throughout its lifetime—for example, changes in the product viscosity.

11.4.1.2 Plume Geometry

The plume geometry is the shape of the plume when looking at the spray from the side as the product is emitted from the device. Like spray pattern testing, plume geometry is used to characterize the performance of the spray pump using a nonimpaction laser sheet-based instrument. Unlike spray pattern testing, however, plume geometry testing is generally only performed during the development stages of the product. The plume geometry is characterized using the following parameters:

- Plume Angle—the angle created at the base of the plume as the product is emitted from the actuator orifice
- Plume Width—the width of the plume at a specified distance from the actuator tip

Table 11.4 Product-specific tests

Test	Comments	Development tests	Bulk	Finished product	Stability
Spray pattern	Metered dose nasal sprays and pressurized aerosols for oral inhalation; sublingual sprays			X	X
Plume geometry	Metered dose nasal sprays and pressurized aerosols for oral inhalation; sublingual sprays	X		X	X
Spray content uniformity/delivered dose uniformity	Metered dose nasal sprays and pressurized aerosols; sublingual sprays			X	X
Droplet size distribution	Metered dose nasal sprays and pressurized aerosols for oral inhalation; sublingual sprays			X	X
Dose weight	Metered dose nasal sprays and pressurized aerosols; sublingual sprays			X	X
Priming/repriming	Multidose metered dose nasal sprays and pressurized aerosols; sublingual sprays	X		X	X
Number of doses	Multidose metered dose nasal sprays and pressurized aerosols; sublingual sprays	X		X	X
Tail-off characteristics	Multidose metered dose nasal sprays and pressurized aerosols; sublingual sprays	X		X	X
Container pressure	Pressurized aerosols fitted with continuous valves			X	X
Delivery rate/delivered amount	Pressurized aerosols fitted with continuous valves			X	X

Test	Pressurized aerosols fitted with continuous valves	Pressurized aerosols for oral inhalation	Emulsions and suspensions	Ointments and sticks	Ophthalmic and otic solutions and suspensions	Suspensions	Ointments and sticks	Ointments and sticks	Ophthalmic ointments
Leakage test	X	X							
Aerodynamic size distribution		X							
Centrifugation			X						
Wick test				X					
Drop weight					X				
Resuspendability						X			
Penetration (viscosity)							X		
Drop point/congealing temperature								X	
Metal particles									X

The size and shape of the actuator orifice, pump design, size of metering chamber (for multidose devices), and formulation characteristics (e.g., viscosity [18]) can also affect the plume geometry.

This testing is performed throughout the lifetime of the product, to determine any changes within the product on storage—for example, changes in the product viscosity.

11.4.1.3 Spray Content Uniformity/Delivered Dose Uniformity (USP<601>)

This testing measures the amount of active ingredient contained within individual sprays across the lifetime of a single device (for multidose sprays), and between devices produced throughout the course of a packaging run (e.g., from the beginning, middle, and end of the filling process). The spray device is actuated using controlled parameters (force, speed, acceleration, stroke length), the emitted spray is captured, and the content of active ingredient (or other functional ingredient) is assayed using the normal product assay method (e.g., HPLC). This testing is performed on the finished product as a routine quality control release test—to ensure that the dosage per actuation (1) complies with the stated label claim, (2) does not change from one actuation to the next, and (3) does not vary from one device to the next. Testing is also performed on storage to determine any changes within the formulation throughout its lifetime—for example, degradation of the active ingredient, loss of preservative, or other functional ingredient.

11.4.1.4 Droplet Size Distribution

This testing is performed using laser diffraction, and characterizes the droplet size within the plume in terms of cumulative volume distributions:

- Dv_{10}, Dv_{50}, Dv_{90}—the volume median diameter (or Dv_{50}) value indicates that half of the spray volume is contained in droplets that are smaller than this value, and half is contained in droplets that are larger than this value. Similarly, the Dv_{10} and Dv_{90} values indicate that 10 and 90%, respectively, of the spray volume is contained in droplets that are smaller than these values
- Span—the span quantifies the spread of the droplet size distribution and is calculated by the following equation: $(Dv_{90} - Dv_{10})/Dv_{50}$
- %<10 μm—the percentage of droplets that are less than 10 μm in diameter

For nasal sprays, droplets that are excessively large ($Dv_{90} > 300$ μm) will have a tendency to either drip out of the nose [19], or hit the back of the throat and be swallowed. The value for %<10 μm provides a risk estimate of small droplets that may be deposited into the lung. Neither of these situations is the intended route of administration and will, therefore, have an impact on the effectiveness of the delivered dose.

Droplet size distribution testing is performed on the finished product as a quality control release test, and on storage to determine any changes within the product throughout its lifetime—for example, changes that will affect the ease of atomization of the formulation (viscosity, surface tension). This testing can also be used during formulation development to assess the impact of different ingredients and their concentrations (e.g., thickeners, muco-adhesives, surfactants) [18].

11.4.1.5 Dose Weight

Dose weight (or pump delivery) testing is performed to assess pump-to-pump reproducibility both within a given product batch, and between different batches of product. The amount of product dispensed by the device will have a direct impact on the efficacy (and safety, in the case of overdelivery) of the product.

Dose weight testing is used both as a finished product quality control release test, and on storage to determine any changes within the product throughout its lifetime—for example, interactions between the formulation and the device may cause changes in the device components that affect the amount of expelled dose. The acceptance criteria in USP<698> (Delivered Volume) may be used as a basis for this testing.

11.4.1.6 Priming/Repriming

Typically, multidose metered spray devices have a dip tube going from the actuator into the product. This is to enable liquid product to be delivered when the device is used in an upright orientation. The device will need priming by actuating several times (to fill the dip tube) before the full dose is emitted. The number of actuations needed to prime the device is obtained by measuring the emitted dose weight until the full (label claim) dose is achieved.

During normal use, these devices will slowly "lose" their prime. That is, the product in the dip tube will slowly run back out into the product container. Repriming exercises are performed by priming the device, storing it in a vertical orientation for a given period of time, then reactuating it to assess whether the first actuation delivers the full dose. The time period after which the device needs additional priming actuations to deliver the full dose, and the number of actuations needed to reprime, are specified in the user instructions.

Priming and repriming exercises are performed during development of the formulation/device combination. This testing might be repeated periodically to check compliance of different batches of product with the guidelines listed in the user instructions. In addition to determining the emitted dose weight, the amount of drug per actuation might also be determined—particularly in the case of suspensions, in which the active ingredient might sediment out from the primed dip tube.

11.4.1.7 Number of Doses

The number of doses contained in a multidose device will depend on the actuation volume, the priming needs, and the container fill weight. This testing is performed by recording the weight of each actuation until the container is exhausted—including the priming actuations. The number of expelled doses is typically determined during development, and is usually listed in the user instructions. The testing might be repeated periodically to check compliance with the user instructions for different batches of product.

11.4.1.8 Tail-Off Characteristics

For multidose spray devices, after the stated number of doses has been expelled some product will remain in the container. The profile of the sprays (spray content uniformity, droplet size distribution) should be determined after the labeled number of doses have been expelled until the container is exhausted. This testing is typically performed during development as the tail-off characteristics are influenced by the pump design, container geometry, and formulation.

11.4.2 Tests Performed on Pressurized Aerosols Fitted with Continuous Valves (e.g., Topical Aerosols) (USP<601>)

11.4.2.1 Container Pressure

This test determines the pressure within the container at a given temperature (typically 25 °C). The container is allowed to equilibrate to the test temperature in a water bath, a pressure gauge is placed on the valve stem, and the valve is actuated so that it is fully open. This test is used as a routine quality release test, as well as on storage—to check for any pressure drop due to, for example, package or valve integrity problems.

11.4.2.2 Delivery Rate/Delivered Amount

This test determines the amount of product dispensed in a given time, and also the total amount dispensed. The container is first primed then allowed to equilibrate to the set temperature (typically 25 °C) in a water bath. The delivery rate is the amount of product dispensed in 5 s, measured by weight loss from the container. To determine the delivered amount, the container is again allowed to equilibrate to the set temperature, and then actuated for a further 5 s. This is repeated until the container is exhausted, the delivered amount being the total weight loss from the container. This testing is used as a routine quality release test, as well as on storage—to check for any changes (e.g., in pressure, the formulation, or the valve) with time. It is important to allow sufficient time between actuations for the product to equilibrate to the set temperature, as cooling will occur (leading to a decrease in the internal pressure) as the product is dispensed.

11.4.2.3 Leakage Test

This test measures the weight loss from the container at a set temperature (typically 25 °C) over a period of not less than three days. This value is then used to determine the leakage rate per year. This testing is used as a routine quality release test, as well as on storage—to check for any changes (e.g., valve integrity problems) with time.

11.4.3 Aerodynamic Size Distribution

The aerodynamic particle or droplet size of solution or suspension inhalation spray drug products will determine whereabouts the product will be deposited within the lungs. It is generally considered that larger particles, with a mass median aerodynamic diameter higher than 10 μm, are deposited in the oropharynx, those measuring between 5 and 10 μm in the central airways, and those from 0.5 to 5 μm in the small airways and alveoli [20]. The optimum aerodynamic particle/droplet size distribution for most oral inhalation products has generally been recognized as being in the range 1–5 μm [17]. It is important, therefore, to determine this parameter as a quality release test, and to monitor any changes on storage.

This testing is performed by cascade impaction [[17], USP<601>]. The spray is actuated into an air stream through an induction port. The particles and/or droplets are fractionated and collected through serial multistage impactions—based on the particle or droplet inertia. The spray collected on each stage of the impactor is then extracted and assayed to give both the size distribution and a total mass balance. As the aerodynamic size measurement is affected by parameters such as impactor design and airflow, these are specified within the test methods.

11.4.4 Stress Tests (Section 7.4.5.1)

11.4.4.1 Centrifugation

Centrifugation testing is used to expose emulsions and suspensions to extreme gravitational forces, to assess formulation stability by accelerating possible phase separation and sedimentation effects. This testing is a quick way of comparing the stabilizing effect of different formulation ingredients (e.g., thickeners, gums), or different processing parameters. Typical centrifugation settings are 3500 rpm for 30 min.

11.4.4.2 Wick Test

Syneresis is an example of formulation instability and is caused by the separation of liquid from the bulk formulation. This can be problematic for ointments and sticks, in which the lighter fractions of the petrolatum base separate out and give the product an oily appearance—even to the extent of oil floating on the top of the formulation. To evaluate, a small sample of the formulation is applied to a piece of filter paper to assess the potential for liquid migrating out of the bulk. This testing can also be used as a way of comparing the stabilizing effect of different formulation ingredients (e.g., different grades of petrolatum), or different processing parameters (e.g., different levels of mixing shear, different mixing temperatures).

11.4.5 Drop Weight

In the case of ophthalmic and otic solutions and suspensions that are delivered from a dropper bottle, the weight of the droplet gives a measure of the delivered dose. If the droplet is too small, the therapeutic dose will not be delivered. If the droplet is too large, overdosing and safety will be a concern. The weight (or size) of the droplet will be affected by the formulation viscosity, and the dropper tip/orifice geometry. Weighing a given number of expressed droplets allows the dose per drop to be determined. Performing this testing during development will allow the correct dropper tip to be selected for a given target dose. This testing can also be used for routine quality release of the product (to ensure lot to lot repeatability), and on storage (to assess formulation changes over time).

11.4.6 Resuspendability

Over time, suspended particles have a tendency to sediment out from suspension formulations—this can be relatively quick (necessitating continual mixing/agitation during processing and filling), or on a much longer time scale (on storage of the formulation). Microscopic examination can provide information on the presence of large particles, changes in particle morphology, extent of agglomerates, and crystal growth. Resuspendability testing evaluates the effect of storage time and conditions on particle size distribution and the ability to homogeneously resuspend the particles by shaking.

Automated equipment is available to shake containers at a fixed displacement/force prior to dispensing or sampling the product for active ingredient content—for example, the integrated Indizo® system (Proveris Scientific Corporation, Marlborough, MA, USA) for metered dose nasal sprays and pressurized inhalers.

11.4.7 Penetration

Penetration testing (per ASTM D127) is used to give a measure of the hardness or consistency of ointment and stick formulations, and is sometimes performed on these formulations in place of viscosity testing. The penetration measurement is the depth (in tenths of a millimeter) to which a standard penetrant such as a cone or a needle sinks into a semisolid substance under defined conditions of sample size, penetrant weight, geometry, and time. The softer the sample is, the deeper the penetrant will sink into the sample and thus the higher the penetration number will be [21].

The ease of processing (e.g., pumping, filling) and skin feel on application will be affected by the penetration value of the formulation. As with viscosity measurements, monitoring penetration at all stages throughout the product lifetime can provide valuable information relating to formulation development, the manufacturing process, batch to batch variability of ingredients and the product, and stability over time.

11.4.8 Drop Point/Congealing Temperature

When measuring the drop point [[22], (USP<741>)], samples are heated until they transform from a solid to a liquid state. The dropping point is the temperature at which the first drop of a substance falls from a cup under defined test conditions. A related measurement, the softening point, is the temperature at which a substance has flowed a certain distance under defined test conditions. The dropping and softening points are mainly used in quality control (to monitor batch to batch variability, and the effect of storage conditions), but can also be valuable in research and development (for determining processing temperatures).

Conversely, the congealing temperature is the temperature at which a substance passes from the liquid to the solid state upon cooling (USP<651>). The sample is melted at a temperature not exceeding 20°C above its expected congealing point, poured into a test tube, stirred continuously, and the temperature recorded every 30 s. When the temperature stops falling (becomes constant or starts to rise slightly), stirring is stopped. Temperature recording is continued every 30 s for at least 3 min after the temperature again begins to fall. The congealing temperature is the point of inflection, or a maximum, in the temperature–time curve.

Although these tests are typically performed on ingredients (waxes, pertolatums, greases), they can also be applied to certain wax- or petrolatum-based semisolid formulations (e.g., ointments and sticks). A melting transition may be instantaneous for a highly pure ingredient, but usually a range is observed from the beginning to the end of the melting process for mixtures of ingredients or formulations.

11.4.9 Metal Particles

This testing is performed to determine the number and size of discrete metal particles in ophthalmic ointments (USP<751>). The ointment is squeezed into a covered Petri dish, heated to 85 °C for 2 h (to ensure that the sample is fully fluid), then cooled to room temperature to solidify. The covers are removed, the Petri dish inverted on the stage of a suitable microscope, and the bottom of the sample examined at 30× magnification with an eyepiece micrometer disk for the presence of metal particles. The number of metal particles that are 50 μm or larger in any dimension is recorded. USP<751> contains more details on the test method, and the acceptable limit for metal particles.

This testing is performed on the finished product as a quality release test. Metal particles, depending on the size and number, could pose a safety risk to the user due to the possibility of the eye becoming scratched or the particle becoming embedded in the eyeball.

REFERENCES

[1] The Tintometer group, www.lovibondcolour.com.
[2] US Pharmacopeia.
[3] Packaging Technologies and Inspection (PTI). Tuckahoe, NY, USA, www.ptiusa.com.
[4] Uson LP. Houston, TX, USA, www.uson.com.
[5] Mocon®. Minneapolis, MN, USA, www.mocon.com.
[6] Malvern Instruments Ltd. Malvern, Worcestershire, UK, www.malvern.com.
[7] rap.ID Particle Systems GmbH. Berlin, Germany, www.rap-id.com.
[8] Diffusion testing fundamentals, www.permegear.com.
[9] 2014 ISTA resource book. International Safe Transit Association (ISTA), www.ista.org.
[10] TA Instruments. New Castle, DE, USA, www.tainstruments.com.

[11] Kauffman JS. Identification and risk-assessment of extractables and leachables. Pharm Technol 2006:s14–22. (Analytical Methods) www.pharmtech.com.

[12] Pace Analytical Life Sciences, LLC. Oakdale, MN, USA, www.pacelifesciences.com.

[13] Intertek Group plc. Houston, TX, USA, www.intertek.com.

[14] Eurofins Lancaster Laboratories. Lancaster, PA, USA, www.lancasterlabs.com.

[15] SGS Life Science Services, a division of SGS North America Inc. Lincolnshire, IL, USA and Fairfield, NJ, USA, www.sgs.com/en/Life-Sciences.

[16] What is Endotoxin? www.lifetechnologies.com.

[17] US FDA. Guidance for industry: nasal spray and inhalation solution, suspension, and spray drug products – chemistry, manufacturing, and controls documentation. July 2002.

[18] Kulkarni VS, Shaw C, Smith M, Brunotte J. Characterization of plumes of nasal spray formulations containing mucoadhesive agents sprayed from different types of device. In: Poster presentation, AAPS annual meeting, San Antonio, Texas. November 2013.

[19] Kulkarni V, Shaw C. Formulation and characterization of nasal sprays. Inhalation June 2012:10–5.

[20] Tena AF, Clara PC. Deposition of inhaled particles in the lungs. Arch Bronconeumologia 2012;48(07):240–6.

[21] Goldenberg WS, Shah R. A useful tool for the determination of consistency in semi-solid substances. New York: Koehler Instrument Company, Inc., www.koehlerinstrument.com.

[22] Excellence dropping point systems. Dropping and softening point determination. Schwerzenbach, Switzerland: Mettler-Toledo AG, Analytical, www.mt.com.

CHAPTER 12

An Overview of Regulatory Aspects for Pharmaceutical Semisolid Dosages

Contents

Although the regulatory landscape for pharmaceutical semisolid products is well established, the regulations are under constant review to improve the quality, safety, and efficacy of products both under development and in the marketplace. This chapter highlights some of the development/quality guidance in place at the time of writing and, more importantly, provides references to where the most up-to-date compliance information can be obtained.

12.1 QUALITY DATA [1]

This section provides a brief summary of the quality data that need to be generated (or considered) for regulatory submissions of pharmaceutical semisolid formulations, at the time of writing.

Essential Chemistry for Formulators of Semisolid and Liquid Dosages
http://dx.doi.org/10.1016/B978-0-12-801024-2.00012-1

12.1.1 Stability

- Stability testing: the regulatory submission should include long-term stability data on three production batches of the formulation packaged in the container closure system as proposed for marketing. Data at accelerated and intermediate (when applicable) conditions should be provided. The climatic zone where the product is to be marketed and the permeability of the packaging are taken into consideration when determining the stability storage conditions (temperature and humidity). Different batches of the drug substance should be used, if possible. Two of the three batches should be at least pilot-scale batches (no less than one-tenth commercial scale), and the third one can be smaller.

- Photostability: photostability testing is carried out on a single batch of material during the development phase, and then the photostability characteristics are confirmed on additional batch(es) of the product when the final packaging is defined. The testing is carried out in a sequential manner starting with the fully exposed product, then progressing to the product in the immediate pack, and then in the marketing pack.

- Bracketing and matrixing: bracketing and matrixing are examples of reduced stability study designs in which every product combination (e.g., strength, package size) is not tested at every time point. When multiple formulation strengths, container sizes, and/or fills exist, bracketing can be used such that only samples at the extremes are tested. For example, bracketing can be applied to formulations with different active ingredient strengths, provided that the formulation excipients remain the same. Similarly, for a given container closure system, bracketing can be applied where either the container size or fill varies, provided the other remains constant. However, if both container size and fill vary, the characteristics of the container closure system that may affect product stability (e.g., container wall thickness, closure geometry, surface area to volume ratio, headspace) must be taken into account when selecting the extremes for testing. Matrixing involves testing a subset of the total number of batches at a specified time point. At a subsequent time point, another subset of batches would be tested. This approach assumes that the stability of each subset of batches represents all samples. Matrixing can be applied, for example, to different formulation strengths in which the relative amounts of drug substance and excipients change or in which different excipients are used, or to different container closure systems. When a secondary packaging system contributes to the stability of the drug product, matrixing can be performed across the packaging systems. Each storage condition should be treated separately under its own matrixing design, and matrixing should not be performed across test attributes.

- Data analysis: when applicable, statistical methods should be employed to analyze the long-term primary stability data. The purpose of this analysis is to establish, with a high degree of confidence, a shelf life throughout which the drug product will remain within acceptance criteria when stored under specified conditions. Regression analysis, with 95% confidence limits, can be used to evaluate the stability data of a quantitative attribute as part of establishing a shelf life.

12.1.2 Analytical Validation

The objective of validation of an analytical procedure is to demonstrate that it is suitable for its intended purpose. The International Conference on Harmonization (ICH) Q2(R1) discusses the validation of the four most common types of analytical procedures: identification tests; quantitative tests for impurities' content; limit tests for the control of impurities; and quantitative tests for the active ingredient or other functional ingredients (e.g., preservatives). The validation characteristics to be considered, depending on the type of test, are: accuracy; precision (to include repeatability and intermediate precision); specificity; detection limit; quantitation limit; linearity; and range. Robustness testing is generally performed during the method development.

12.1.3 Impurities

Impurities in pharmaceutical products can be categorized as organic, inorganic, metallic, or residual solvents [1], and their presence can pose a safety concern for the user. These impurities could originate from materials used during the manufacture of the drug substance or excipients and be transferred through to the finished product, or formed in the finished product formulation through the combination of the drug substance and excipients and/or container closure. Threshold values should be determined based on the maximum daily dose of the drug substance administered in the drug product according to ICH Q3B(R2).

Metallic (elemental) impurities might arise from interactions with processing equipment, and do not provide any therapeutic benefit to the patient. Because these impurities could be harmful to the patient, they should be minimized.

Residual solvents in pharmaceuticals are defined in Q3C(R5) as organic volatile chemicals that are used or produced in the manufacture of drug substances or excipients, or in the preparation of drug products. Whenever possible, solvents that are less toxic should be used.

12.1.4 Biotechnological Products

- Viral safety: the risk of viral contamination is common to all biotechnology products derived from cell lines, and could have serious clinical consequences. This contamination can arise from the contamination of the source cell lines themselves (cell substrates), or from introduction of the virus during production. ICH Q5A(R1) notes that the viral safety of biotechnology can be controlled through: selecting and testing cell lines and other raw materials for the absence of undesirable viruses which may be infectious and/or pathogenic for humans; assessing the capacity of the production processes to clear infectious viruses; and testing the product at appropriate steps of production for absence of contaminating infectious viruses.

- Coding sequence: ICH Q5B tells us that the genetic sequence of recombinant proteins produced in living cells can undergo mutations that could alter the properties of the protein with potential adverse consequences to patients. The purpose of analyzing the expression construct is to establish that the correct coding sequence of the product has been incorporated into the host cell, and is maintained through to the end of production. Analytical data derived from both nucleic acid analysis and evaluation of the final purified protein can be used to assess the amino acid sequence of the protein as well as structural features of the expressed protein.

- Stability testing: products containing proteins and/or polypeptides are particularly sensitive to factors such as temperature changes, oxidation, light, ionic content, and processing shear. The maintenance of molecular conformation and biological activity should be demonstrated throughout the shelf life of the product. Physicochemical, biochemical, and immunochemical methods should be used, as appropriate, for the analysis of the molecular entity and the quantitative detection of degradation products as part of the stability program.

- Production and characterization: the quality and safety of biotechnological/biological products can be affected by the properties and handling of the cell line. Documentation which describes the history of the cell line, as well as any parental cell line from which it was totally or partially derived, should be provided in the regulatory submission. This should include the origin, source, cultivation history, and cell banking systems and procedures. Testing should be performed to confirm the identity, purity, and suitability of the cell line for manufacturing use, and its stability.

- Effect of manufacturing changes: changes to the manufacturing process are sometimes made during the lifetime of biotechnological/biological products (e.g., process improvement, scale-up, to improve product stability, and to comply with regulatory changes). Comparability testing is performed pre- and postchange to demonstrate that the drug product is not adversely impacted in terms öf quality, safety, and efficacy.

12.1.5 Specifications

Specifications establish the criteria to which a new drug product, its chemical components, and its container closure should conform to be considered acceptable for its intended use. ICH Q6A tells us that specifications constitute one element of the quality controls applied to the development and manufacture of a drug product. Other elements include their design, development, in-process controls, Good Manufacturing Practice controls, and process validation. Data accumulated during the development of the drug product should form the basis for the setting of specifications, along with a reasonable range of expected analytical and manufacturing variability. In the case in which drug products are covered by pharmacopoeial monographs, the listed procedures and/or acceptance criteria should be used.

The amount of testing performed may vary depending on the stage or the phase of the product in the manufacturing process (e.g., in-process or finished product) or the age of the product in relation to its shelf-life (e.g., full testing on product release, and reduced testing at certain storage intervals). The acceptance criteria for release and shelf-life specifications might also be different—more restrictive criteria might be used for the release of a drug product than are applied to the shelf life.

For biotechnological/biological products, specifications should focus on those molecular and biological characteristics found to be useful in ensuring the safety and efficacy of the product.

12.1.6 Pharmaceutical Development

ICH Q8(R2) tells us that the aim of pharmaceutical development is to design a quality product and its manufacturing process to consistently deliver the intended performance of the product. The information presented in a regulatory submission should include:

- the physicochemical and biological properties of the drug substance that can influence the performance of the drug product (e.g., respirable fraction of an inhaled product) and its manufacturability

- the excipients chosen, their concentration, and the characteristics that can influence the drug product performance (e.g., stability, bioavailability) or manufacturability
- a summary describing the development of the formulation, including identification of critical quality attributes of the drug product, taking into consideration intended usage and route of administration
- manufacturing process development, including process control strategies
- choice of container closure system and materials
- microbiological attributes (e.g., choice and level of preservatives).

12.1.7 Quality Risk Management

Quality risk management can be applied to all aspects of pharmaceutical quality—including development, manufacturing, inspection, and distribution of drug, biological and biotechnological products. It is a proactive means of systematically identifying and controlling potential quality issues during development and manufacturing. Tools such as Failure Mode Effects Analysis (FMEA), Failure Mode Effects and Criticality Analysis (FMECA), Fault Tree Analysis (FTA), Hazard Analysis and Critical Control Points (HACCP), Hazard Operability Analysis (HAZOP), Preliminary Hazard Analysis (PHA), and Risk Ranking and Filtering are used to identify and assess possible risks. Once the risks have been determined, manufacturing process controls (including proposed batch records, and container closure specifications and testing) can be used as an ongoing means of control.

12.2 TYPES OF REGULATORY SUBMISSIONS

This section provides a brief summary of the different categories of regulatory submissions of pharmaceutical semisolid formulations in the USA, as described by the U.S. Food and Drug Administration (FDA) [2].

- Investigational New Drug (IND): current Federal law requires that a drug be the subject of an approved marketing application before it is transported or distributed into interstate commerce. Because the drug product being developed has not been approved, a sponsor must seek an exemption from that legal requirement. The IND is the means through which the sponsor technically obtains this exemption from the FDA. An IND is not required for drug product being used for an Abbreviated New Drug Application (ANDA) bioequivalence study.

- New Drug Application (NDA): the NDA application is the vehicle through which drug sponsors formally propose that the FDA approve a new pharmaceutical for sale and marketing in the US. The data gathered during the animal studies and human clinical trials under an IND become part of the NDA. The NDA submission should demonstrate that: (1) the drug is safe and effective in its proposed use(s); (2) the drug's proposed labeling (package insert) is appropriate; and (3) the methods used in manufacturing the drug and the controls used to maintain the drug's quality are adequate to preserve the drug's identity, strength, quality, and purity.
- ANDA: an ANDA is the regulatory application covering generic drug products. A generic drug product is one that is comparable to an innovator drug product in terms of dosage form, strength, route of administration, quality, performance characteristics, and intended use. Generic drug applications are termed "abbreviated" because they are generally not required to include preclinical (animal) and clinical (human) data to establish safety and effectiveness. Instead, generic applicants must scientifically demonstrate that their product is bioequivalent (i.e., performs in the same manner as the innovator drug ["Reference Listed Drug"]).
- 505(b)(2): 505(b)(2) applications are submitted in cases in which some part of the data necessary for approval is derived from studies not conducted by or for the applicant, and to which the applicant has not obtained a right of reference. Examples are: (1) a New Chemical Entity (NCE), in which the data is derived from published literature studies; and (2) changes to approved drugs (including changes in dosage form, strength, route of administration, or substitution of an active ingredient in a combination product), in which bioavailablity or bioequivalence information is presented.
- Over-the-counter (OTC): OTC (nonprescription) drug products are defined as drugs that are safe and effective for use by the general public without seeking treatment by a health professional. OTC drug monographs exist for different product categories (e.g., analgesics or antacids), and constitute a standard for acceptable active ingredients, labeling, and, in some cases, testing prior to marketing. Once a final monograph is implemented, companies can manufacture and market an OTC product without the need for FDA preapproval.
- Animal Drug Products: similar to above, New Animal Drug Applications (NADA) and Abbreviated New Animal Drug Applications (ANADA) cover regulatory submissions for animal drug products.

12.3 RESOURCES—ICH

The International Conference on Harmonization of Technical Requirements for Registration of Pharmaceuticals for Human Use (ICH) [1] was founded in 1990 to bring together the regulatory authorities and pharmaceutical industry of Europe, Japan, and the US to discuss scientific and technical aspects of drug registration. Its mission is to achieve greater harmonization to ensure that safe, effective, and high-quality medicines are developed and registered in the most resource-efficient manner.

The guidelines developed by ICH fall into four categories: Quality, Safety, Efficacy, and Multidisciplinary Guidelines.

12.3.1 Quality Guidelines Relating to Drug Products

* Q1A(R2) "Stability Testing of New Drug Substances and Products"
  This guideline defines the stability data package (including temperature, humidity, test intervals, and trial duration) for a new drug substance or drug product that is sufficient for a registration application within the three regions of the European Community (EC), Japan, and the United States.
* Q1B "Stability Testing: Photostability Testing of New Drug Substances and Products"
  This document provides guidance on the basic testing protocol required to evaluate the light sensitivity and stability of new drugs and products. Information on suitable light sources is also presented.
* Q1C "Stability Testing for New Dosage Forms"
  This guideline relates to new dosage forms of already approved medicines (i.e., a different pharmaceutical product type, but containing the same active substance), and defines the circumstances under which reduced stability data can be accepted. The document supplements the main stability guideline (Q1A[R2]).
* Q1D "Bracketing and Matrixing Designs for Stability Testing of New Drug Substances and Products"
  This guideline defines general principles in which bracketing or matrixing can be applied to reduce the amount of stability testing, and provides examples of bracketing and matrixing designs.
* Q1E "Evaluation of Stability Data"
  This document provides examples of statistical approaches to analyzing stability data in order to propose a retest period or shelf life in a registration application. It also describes when and how extrapolation of stability data may be appropriate.

- Q2(R1) "Validation of Analytical Procedures: Text and Methodology"
  This document identifies the parameters for validation for a variety of analytical methods. It includes information relating to the actual experimental data required, along with its statistical interpretation.
- Q3A(R2) "Impurities in New Drug Substances"
  This guideline addresses the chemistry and safety aspects of impurities in new drug substances produced by chemical syntheses.
- Q3B(R2) "Impurities in New Drug Products"
  This guideline provides advice regarding the content and qualification of impurities in products containing new, chemically synthesized drug substances. It deals with those impurities which might arise as degradation products of the drug substance, or arising from interactions between drug substance and excipients or components of packaging materials.
- Q3C(R5) "Impurities: Guideline for Residual Solvents"
  This guideline recommends acceptable amounts for residual solvents in pharmaceuticals in relation to the safety of the patient.
- Q3D "Impurities: Guideline for Elemental Impurities"
  This guidance is still in draft at the time of writing. It proposes evaluating the toxicity data for potential elemental impurities, establishing a Permitted Daily Exposure (PDE) for each element of toxicological concern, and developing controls designed to limit the inclusion of elemental impurities in drug products to levels at or below the PDE.
- Q5A(R1) "Viral Safety Evaluation of Biotechnology Products Derived from Cell Lines of Human or Animal Origin"
  This document provides a framework for virus testing, experiments for evaluating virus clearance, and the design of viral tests and clearance evaluation studies.
- Q5B "Analysis of the Expression Construct in Cells Used for Production of r-DNA Derived Protein Products"
  This guideline provides advice on the types of information that are valuable in assessing the structure of the expression construct used to produce recombinant DNA-derived proteins.
- Q5C "Stability Testing of Biotechnological/Biological Products"
  This document addresses stability test procedures for products in which the active components are proteins and/or polypeptides.
- Q5D "Derivation and Characterisation of Cell Substrates Used for Production of Biotechnological/Biological Products"
  This document provides guidance on standards applicable for the preparation and characterization of cell lines and cell banks used for the production of biotechnological/biological products.

- Q5E "Comparability of Biotechnological/Biological Products Subject to Changes in their Manufacturing Process"
This document provides guidance for assessing the comparability of biotechnological/biological products before and after changes are made in the manufacturing process.
- Q6A "Specifications: Test Procedures and Acceptance Criteria for New Drug Substances and New Drug Products: Chemical Substances"
This guideline addresses the process of selecting tests and methods and setting specifications for the testing of drug substances and dosage forms,
- Q6B "Specifications: Test Procedures and Acceptance Criteria for Bio-technological/Biological Products"
This guideline addresses selecting tests and methods, and setting specifications for proteins and polypeptides which are derived from recombinant or nonrecombinant cell cultures.
- Q8(R2) "Pharmaceutical Development"
This document provides guidance on the information required in the Pharmaceutical Development section of a drug product submission.
- Q9 "Quality Risk Management"
This guideline provides principles and examples of tools of quality risk management.

12.3.2 Other Quality Guidelines

- Q4 "Pharmacopoeias"
- Q4A "Pharmacopoeial Harmonization"
- Q4B "Evaluation and Recommendation of Pharmacopoeial Texts for Use in the ICH Regions"
- Q7 "Good Manufacturing Practice Guide for Active Pharmaceutical Ingredients"
- Q10 "Pharmaceutical Quality System"
- Q11 "Development and Manufacture of Drug Substances."

12.3.3 Safety Guidelines

- S1A "Need for Carcinogenicity Studies of Pharmaceuticals"
This guideline defines the circumstances under which it is necessary to undertake carcinogenicity studies on new drugs.
- S1B "Testing for Carcinogenicity of Pharmaceuticals"
This document provides guidance on the need to carry out carcinogenicity studies in mice and rats, as well as alternative testing procedures.

- S1C(R2) "Dose Selection for Carcinogenicity Studies of Pharmaceuticals"
  This document addresses the criteria for the selection of the high dose to be used in carcinogenicity studies on new therapeutic agents.
- S2(R1) "Guidance on Genotoxicity Testing and Data Interpretation for Pharmaceuticals Intended for Human Use"
  This document outlines genetic toxicology tests for predicting potential carcinogenic effects that have their basis in changes in the genetic material.
- S3A "Note for Guidance on Toxicokinetics: The Assessment of Systemic Exposure in Toxicity Studies"
  This document provides guidance on developing test strategies in toxicokinetics and integrating pharmacokinetics into toxicity testing.
- S3B "Pharmacokinetics: Guidance for Repeated Dose Tissue Distribution Studies"
  This document provides guidance on circumstances when repeated dose tissue distribution studies should be considered.
- S4 "Duration of Chronic Toxicity Testing in Animals (Rodent and Non Rodent Toxicity Testing)"
  This document provides guidance on chronic toxicity testing in rodents and non-rodents as part of the safety evaluation of a medicinal product.
- S5(R2) "Detection of Toxicity to Reproduction for Medicinal Products and Toxicity to Male Fertility"
  This document provides guidance on tests for reproductive toxicity and the periods of treatment to be used in animals.
- S6(R1) "Preclinical Safety Evaluation of Biotechnology-Derived Pharmaceuticals"
  This document covers the preclinical safety testing requirements for biotechnological products—including the use of animal models of disease, determination of when genotoxicity assays and carcinogenicity studies should be performed, and the impact of antibody formation on the duration of toxicology studies.
- S7A "Safety Pharmacology Studies for Human Pharmaceuticals"
  This document addresses which safety pharmacology studies are needed before initiation of Phase 1 clinical studies as well as information needed for marketing.
- S7B "The Non-Clinical Evaluation of the Potential for Delayed Ventricular Repolarization (QT Interval Prolongation) by Human Pharmaceuticals"
  This guideline describes a nonclinical testing strategy for assessing the potential of a test substance to delay ventricular repolarization.

- S8 "Immunotoxicity Studies for Human Pharmaceuticals"
  This guideline provides recommendations on nonclinical testing for immunosuppression (i.e., a state of increased susceptibility to infections or the development of tumors) induced by low molecular weight drugs (nonbiologicals).
- S9 "Nonclinical Evaluation of Anticancer Pharmaceuticals"
  This guideline describes the type and timing of nonclinical studies for pharmaceuticals (both small molecule and biotechnology-derived) that are intended to treat cancer in patients with late stage or advanced disease, regardless of the route of administration.
- S10 "Photosafety Evaluation of Pharmaceuticals"
  This guideline provides information on the photosafety assessment of pharmaceuticals to support human clinical trials and marketing authorizations.

12.3.4 Efficacy Guidelines

- E1 "The Extent of Population Exposure to Assess Clinical Safety for Drugs Intended for Long-Term Treatment of Non-Life Threatening Conditions"
  This document gives recommendations on the numbers of patients and duration of exposure for the safety evaluation of drugs intended for the long-term treatment of non-life-threatening conditions.
- E2A "Clinical Safety Data Management: Definitions and Standards for Expedited Reporting"
  This document provides guidance on standard definitions and terminology for clinical safety reporting, and the mechanism for expedited reporting.
- E2C(R2) "Periodic Benefit-Risk Evaluation Report"
  This document gives guidance on the format and content of safety updates to be provided to regulatory authorities after products have been marketed.
- E2D "Post-Approval Safety Data Management: Definitions and Standards for Expedited Reporting"
  This document provides information relating to postapproval safety data management and reporting.
- E2E "Pharmacovigilance Planning"
  This guideline is intended to aid in the planning of postmarketing pharmacovigilance activities for new drugs as additional information becomes available.
- E2F "Development Safety Update Report"
  This document provides guidance on the periodic reporting of safety information on drugs under development (including marketed drugs that are under further study).

- E3 "Structure and Content of Clinical Study Reports"
  This document describes the format and content of a study report that will be acceptable to all regulatory authorities of the ICH regions.
- E4 "Dose-Response Information to Support Drug Registration"
  This document gives guidance on studies to assess the relationship between dose, blood concentration, and clinical response throughout the clinical development of a new drug.
- E5(R1) "Ethnic Factors in the Acceptability of Foreign Clinical Data"
  This document provides guidance on extrapolating foreign clinical data to new regions, including performing bridging studies, to minimize duplication of clinical studies.
- E6(R1) "Good Clinical Practice"
  This document describes the responsibilities and expectations for all participants involved in the conduct of clinical trials (including investigators, monitors, sponsors, and Institutional Review Boards [IRBs]), and covers items such as monitoring, reporting, and archiving of clinical trials.
- E7 "Studies in Support of Special Populations: Geriatrics"
  This document provides guidance on the design and conduct of clinical trials for medicines that are likely to have significant use in the elderly.
- E8 "General Considerations for Clinical Trials"
  This document outlines the general scientific principles for the conduct, performance, and control of clinical trials at all clinical phases of development.
- E9 "Statistical Principles for Clinical Trials"
  This guideline describes biostatistical considerations relating to the design and analysis of clinical trials, in particular clinical trials conducted in the later phases of development—many of which are confirmatory trials of efficacy.
- E10 "Choice of Control Group and Related Issues in Clinical Trials"
  This guideline describes the general principles involved in choosing a control group for clinical trials intended to demonstrate the efficacy of a treatment, and discusses related trial design and conduct issues. It stresses the importance of the choice of the control group in the context of available standard therapies, the adequacy of the evidence to support the chosen design, and ethical considerations.
- E11 "Clinical Investigation of Medicinal Products in the Pediatric Population"
  This document addresses the conduct of safe, efficient, and ethical clinical trials of medicines in pediatric populations.

- E12 "Principles for Clinical Evaluation of New Antihypertensive Drugs"
  This document covers the clinical evaluation of new antihypertensive drugs, and gives consideration to both diastolic and systolic hypertension.
- E14 "The Clinical Evaluation of QT/QTc Interval Prolongation and Proarrhythmic Potential for Non-Antiarrhythmic Drugs"
  This document provides guidance on the design, conduct, analysis, and interpretation of clinical studies to assess the potential of a drug to delay cardiac repolarization.
- E15 "Definitions for Genomic Biomarkers, Pharmacogenomics, Pharmacogenetics, Genomic Data and Sample Coding Categories"
  This guideline contains definitions of key terms in the discipline of pharmacogenomics and pharmacogenetics.
- E16 "Biomarkers Related to Drug or Biotechnology Product Development: Context, Structure and Format of Qualification Submissions"
  This guideline recommends a structure for biomarker qualification applications to support the use of a biomarker during drug or biotechnology product development.

12.3.5 Multidisciplinary Guidelines

This section lists ICH guidelines that do not fit uniquely into one of the Quality, Safety, and Efficacy categories.

- M1 "Medical Dictionary for Regulatory Activities" (MedDRA Terminology)
- M2 "Electronic Standards for the Transfer of Regulatory Information"
- M3(R2) "Guidance on Nonclinical Safety Studies for the Conduct of Human Clinical Trials and Marketing Authorizations for Pharmaceuticals"
- M4 "The Common Technical Document"
- M5 "Data Elements and Standards for Drug Dictionaries"
- M6 "Virus and Gene Therapy Vector Shedding and Transmission"
- M7 "Assessment and Control of DNA Reactive (Mutagenic) Impurities in Pharmaceuticals to Limit Potential Carcinogenic Risk"
- M8 "Electronic Common Technical Document (eCTD)"

12.4 RESOURCES—FDA

The U.S. FDA has produced a great number of guidance documents and literature covering regulatory compliance information for animal and veterinary products, biologics, cosmetics, drug products, medical devices, and tobacco products [2], including topics such as:

- Bioequivalence
- Biopharmaceutics
- Biosimilarity
- Chemistry, Manufacturing, and Controls (CMC)
- Clinical Investigations
- Combination Products
- Current Good Manufacturing Practices (cGMP)
- Drug Safety
- Electronic Submissions
- Generics
- Investigational New Drugs
- Labeling
- Microbiology
- Over-the-Counter Drugs
- Pharmacology/Toxicology

12.5 RESOURCES—WHO

The World Health Organization has produced guidelines to define the core stability data package required for registration of active pharmaceutical ingredients and finished pharmaceutical products [3]. In addition to outlining the selection of batches for testing, the testing frequency, and giving examples of parameters to be tested, the guidelines list suitable long-term testing conditions (temperature and humidity) for WHO Member States based on climatic data.

REFERENCES

[1] www.ich.org.
[2] www.fda.gov.
[3] Stability testing of active pharmaceutical ingredients and finished pharmaceutical products. World Health Organization, WHO Technical Report Series 2009; (Annex 2): No. 953.

INDEX

Note: 'Page numbers followed by "f" indicate figures and "t" indicate tables.'